BINARY DECISION DIAGRAMS AND APPLICATIONS FOR VLSI CAD

THE KLUWER INTERNATIONAL SERIES IN ENGINEERING AND COMPUTER SCIENCE

VLSI, COMPUTER ARCHITECTURE AND DIGITAL SIGNAL PROCESSING

Consulting Editor

Jonathan Allen

Other books in the series:

ROBUSTNESS IN AUTOMATIC SPEECH RECOGNITION, Jean-Claude Junqua, Jean-Paul Haton
ISBN: 0-7923-9646-4

HIGH-PERFORMANCE DIGITAL VLSI CIRCUIT DESIGN, Richard X. Gu, Khaled M. Sharaf, Mohamed I. Elmasry
ISBN: 0-7923-9641-3

LOW POWER DESIGN METHODOLOGIES, Jan M. Rabaey, Massoud Pedram
ISBN: 0-7923-9630-8

MODERN METHODS OF SPEECH PROCESSING, Ravi P. Ramachandran
ISBN: 0-7923-9607-3

LOGIC SYNTHESIS FOR FIELD-PROGRAMMABLE GATE ARRAYS, Rajeev Murgai, Robert K. Brayton
ISBN: 0-7923-9596-4

CODE GENERATION FOR EMBEDDED PROCESSORS, P. Marwedel, G. Goossens
ISBN: 0-7923-9577-8

DIGITAL TIMING MACROMODELING FOR VLSI DESIGN VERIFICATION, Jeong-Taek Kong, David Overhauser
ISBN: 0-7923-9580-8

DIGIT-SERIAL COMPUTATION, Richard Hartley, Keshab K. Parhi
ISBN: 0-7923-9573-5

FORMAL SEMANTICS FOR VHDL, Carlos Delgado Kloos, Peter T. Breuer
ISBN: 0-7923-9552-2

ON OPTIMAL INTERCONNECTIONS FOR VLSI, Andrew B. Kahng, Gabriel Robins
ISBN: 0-7923-9483-6

SIMULATION TECHNIQUES AND SOLUTIONS FOR MIXED-SIGNAL COUPLING IN INTEGRATED CIRCUITS, Nishath K. Verghese, Timothy J. Schmerbeck, David J. Allstot
ISBN: 0-7923-9544-1

MIXED-MODE SIMULATION AND ANALOG MULTILEVEL SIMULATION, Resve Saleh, Shyh-Jye Jou, A. Richard Newton
ISBN: 0-7923-9473-9

CAD FRAMEWORKS: Principles and Architectures, Pieter van der Wolf
ISBN: 0-7923-9501-8

PIPELINED ADAPTIVE DIGITAL FILTERS, Naresh R. Shanbhag, Keshab K. Parhi
ISBN: 0-7923-9463-1

TIMED BOOLEAN FUNCTIONS: A Unified Formalism for Exact Timing Analysis, William K.C. Lam, Robert K. Brayton
ISBN: 0-7923-9454-2

AN ANALOG VLSI SYSTEM FOR STEREOSCIPIC VISION, Misha Mahowald
ISBN: 0-7923-944-5

BINARY DECISION DIAGRAMS AND APPLICATIONS FOR VLSI CAD

by

Shin-ichi Minato

NTT LSI Laboratories
Kanagawa, Japan

KLUWER ACADEMIC PUBLISHERS
Boston / Dordrecht / London

Distributors for North America:
Kluwer Academic Publishers
101 Philip Drive
Assinippi Park
Norwell, Massachusetts 02061 USA

Distributors for all other countries:
Kluwer Academic Publishers Group
Distribution Centre
Post Office Box 322
3300 AH Dordrecht, THE NETHERLANDS

Library of Congress Cataloging-in-Publication Data

A C.I.P. Catalogue record for this book is available
from the Library of Congress.

Printed on acid-free paper.

Printed in the United States of America

CONTENTS

FOREWORD

Symbolic Boolean manipulation using Binary Decision Diagrams (BDDs) has been applied successfully to a wide variety of tasks, particularly in very large scale integration (VLSI) computer-aided design (CAD). The concept of decision graphs as an abstract representation of Boolean functions dates back to early work by Lee and Akers. In the last ten years, BDDs have found widespread use as a concrete data structure for symbolic Boolean manipulation. With BDDs, functions can be constructed, manipulated, and compared by simple and efficient graph algorithms. Since Boolean functions can represent not just digital circuit functions, but also such mathematical domains as sets and relations, a wide variety of CAD problems can be solved using BDDs.

Although I can claim some credit in initiating the use of BDDs for symbolic Boolean manipulation, the state of the art in the field has been advanced by researchers around the world. In particular, the group headed by Prof. Shuzo Yajima at Kyoto University has been the source of many important research results as well as the spawning ground for some of the most productive researchers. Shin-Ichi Minato is a prime example of this successful research environment. While a Master's student at Kyoto University, he and his colleagues developed important refinements to the BDD data structure, including a shared representation, attributed edges, and improved variable ordering techniques. Since joining the NTT LSI Laboratories, Minato has continued to make important research contributions. Perhaps most significant among these is the use of a zero-suppressed reduction rule when using BDDs to represent sparse sets. These "ZBDDs" have proved effective for such tasks as the cube-set manipulations in two-level logic optimization, as well as for representing polynomial expressions.

This book is based on Minato's Phd dissertation and hence focuses on his particular contributions to the field of BDDs. The book provides valuable information for both those who are new to BDDs as well as to long time aficionados. Chapters 1–4 provide a thorough and self-contained background to the area, in part because Minato's contributions have become part of the mainstream practice. Chapters 5–7 present an area for which Minato's contributions have been of central importance, namely the encoding of combinatorial problems as sparse sets and their solution using ZBDDs. Chapters 8–9 carry these ideas beyond Boolean functions, using BDDs to represent polynomial expressions and integer-valued functions. These latter parts provide intriguing insights into how BDDs can be applied outside VLSI CAD.

Randal E. Bryant
Carnegie Mellon University

PREFACE

This book is a revised edition of my doctoral thesis, submitted to Kyoto University, Japan, in 1995. It is mainly a report of my research on the techniques of Boolean function manipulation using *Binary Decision Diagrams (BDDs)* and their applications for VLSI CAD systems. This research was done in Prof. Yajima's Laboratory at Kyoto University from 1987 to 1990 and in NTT LSI Laboratories from 1990 to 1995.

Over the past ten years, since appearance of Bryant's paper in 1986, BDDs have attracted the attention of many researchers because of their suitability for representing Boolean functions. They are now widely used in many practical VLSI CAD systems. I hope that this book can serve as an introduction to BDD techniques and that it presents several new ideas on BDDs and their applications. I expect many computer scientists and engineers will be interested in this book since Boolean function manipulation is a fundamental technique not only in digital system design but also in exporing various problems in computer science.

The contributions of this book are summarized as follows. Chapters 2 through 4 present implementation and utility techniques of BDDs. Currently, there are few books or papers concisely describing the basic method of BDD manipulation, so those chapters can be read as an introduction to BDDs.

Transforming BDDs into another data structure is an important issue. Chapter 5 proposes a fast method for generating compact cube sets from BDDs. Cube set data structure is commonly used in digital system design, and the proposed method will find many applications since the cube sets are also used for representing a kind of knowledge in artificial intelligence or data base systems.

The concept of *Zero-suppressed BDDs (ZBDDs)*, introduced in Chapter 6, is the most important idea in this book. The use of ZBDDs results in the implicit cube set representation described in Chapter 7, and it greatly accelerates multi-level logic synthesis systems and enlarges the scale of the circuits to which these systems are applicable. Chapter 8 provides another application of ZBDDs: the manipulation of arithmetic polynomial formulas. This chapter presents the newest topic, not included in the original doctoral thesis.

Chapter 9 can be read independently from the other chapters. Here we apply the BDD techniques for solving various problems in computer science, such as the 8-queens problem, the traveling salesman problem, and a kind of scheduling problems. We developed an arithmetic Boolean expression manipulator, which is a helpful tool for the research or education related to discrete functions. This program, runs on a SPARC station, is open to the public. Anyone interested in this program can get it from an anonymous FTP server authorized by the *Information Processing Society of Japan*:
eda.kuee.kyoto-u.ac.jp (130.54.29.134) /pub/cad/BemII.

Throughout I have assumed that the reader of this book is a researcher, an engineer, or a student who is familiar with switching theory. Since BDD is a graph-based representation, I have endeavored to provide plentiful illustrations to help the reader's understanding, and I have shown experimental results whenever possible, in order to demonstrate the practical side of my work.

As always, this work would not have been possible without the help of many other people. First I would like to express my sincere appreciation to Professor Shuzo Yajima of Kyoto University for his continuous guidance, interesting suggestions, and encouragement during this research. I would also like to express my thanks to Dr. Nagisa Ishiura of Osaka University, who introduced me to the research field of VLSI CAD and BDDs, and has been giving me invaluable suggestions, accurate criticism, and encouragement throughout this research.

I express my deep gratitude to Professor Randal E. Bryant of Carnegie Mellon University. I started this research with his papers, and I received valuable suggestions and comments from him. I am honored by his writing the Foreword to this book.

I also acknowledge interesting comments that I have received from Professor Hiromi Hiraishi of Kyoto Sangyo University, Professor Hiroto Yasuura of Kyushuu University, and Associate Professor Naofumi Takagi of Nagoya University. I would like to thank Dr. Kiyoharu Hamaguch, Mr. Koich Yasuoka, and Mr. Shoichi Hirose of Kyoto University, Associate Professor Hiroyuki Ochi of Hiroshima City University, Mr. Yasuhito Koumura of Sanyo Corporation, and Kazuya Ioki of IBM Japan Corporation for their interesting discussions and for their collaboration in implementing the BDD program package.

Special thanks are also due to Professor Saburo Muroga of the University of Illinois at Urbana-Champaign and Professor Tsutomu Sasao of Kyushu Institute of Technology for their instructive comments and advice on the publication of this book. I am grateful to Dr. Masahiro Fujita and Mr. Yusuke Matsunaga of Fujitsu Corporation, who started to work on BDDs early. I referred to their papers many times, and had fruitful discussions with them.

I also wish to thank Dr. Olivier Coudert of Synopsys corporation for his discussions on the implicit set representation, which led to the idea of ZBDDs. I would like to thank Professor Gary D. Hachtel and Associate Professor Fabio Somenzi of the University of Colorado at Boulder for their fruitful discussions on the application of BDDs to combinatorial problems. I am also grateful to Professor Robert K. Brayton of the University of Calfornia at Berkeley, Dr. Patrik McGeer of Cadance Berkeley Laboratolies, Dr. Yoshinori Watanabe of Digital Equipment Corporation, and Mr. Yuji Kukimoto of the University of Calfornia at Berkeley for their interesting discussions and suggestions on the techniques of BDD-based logic synthesis.

Thanks are also due to my colleagues in NTT Research Laboratories. I gratefully acknowledge the collaboration with Dr. Toshiaki Miyazaki and Dr. Atsushi Takahara in developing and evaluating the BEM-II program. I am also grateful to Mr. Hiroshi G. Okuno and Mr. Masayuki Yanagiya for their interest in BEM-II applications. I also thank Mr. Noriyuki Takahashi for his helpful comments on the application of ZBDDs for fault simulation.

I wish to express my special gratitude to Mr. Yasuyoshi Sakai, Dr. Osamu Karatsu, Mr. Tamio Hoshino, Mr. Makoto Endo, and all the other members of NTT Advanced LSI Laboratory for giving me an opportunity to write this book and for their advice and encouragement. Thanks are due to all the members of the Professor Yajima's Laboratory for their discussions and helpful supports throughout this research.

Lastly, I thank my parents and my wife for their patience, support, and encouragement.

BINARY DECISION DIAGRAMS AND APPLICATIONS FOR VLSI CAD

1 INTRODUCTION

1.1 BACKGROUND

The manipulation of Boolean functions is a fundamental part of computer science, and many problems in the design and testing of digital systems can be expressed as a sequence of operations on Boolean functions. The recent advance in very large-scale integration (VLSI) technology has caused these problems to grow beyond the scope of manual design procedures, and has resulted in the wide use of computer-aided design (CAD) systems. The performance of these systems greatly depends on the efficiency with which Boolean functions are manipulated, and this is also a very important technique in computer science problems such as artificial intelligence and combinatorics.

A good data structure is a key to efficient Boolean function manipulation. The following basic tasks must be performed efficiently in terms of execution time and memory space.

- Generating Boolean function data, which is the result of a logic operation (such as AND, OR, NOT, and EXOR), for given Boolean functions.
- Checking the tautology or satisfiability of a given Boolean function.
- Finding an assignment of input variables such that a given Boolean function becomes 1, or counting the number of such assignments.

Various methods for representing and manipulating Boolean functions have been developed, and some of the classical methods are *truth tables, parse trees*, and *cube sets*.

Truth tables are suitable for manipulation on computers, especially on recent high-speed vector processors[IYY87] or parallel machines, but they need 2^n bit of memory to represent an n-input function — even a very simple one. A 100-input tautology function, for example, requires 2^{100} bit of truth table. Since an exponential memory requirement leads to an exponential computation time, truth tables are impractical for manipulating Boolean functions with many input variables.

Parse trees for Boolean expressions sometimes give compact representations for functions that have many input variables, and that cannot be represented compactly using truth tables. A drawback to the use of parse trees, however, is that there are many different expressions for a given function. Checking the equivalence of two expressions is very hard because it is an NP problem, although rule-based method for transformation of the Boolean expressions have been developed[LCM89].

Cube sets (also called sum-of-products, PLA forms, covers, or two-level logics) are regarded as a special form of the Boolean expressions with the AND-OR two level structure. They have been extensively studied for many years and have been used to represent Boolean functions on computers. Cube sets sometimes give more compact representation than truth tables, but redundant cubes may appear in logic operations and they have to be reduced in order to check tautology or equivalency. This reduction process is time consuming. Other drawbacks are that NOT operation cannot be performed easily and that the size of cube sets for parity functions become exponential.

Because the classical methods are impractical for large-scale problems, it have been necessary to develop an efficient method for representing practical Boolean functions. The basic concept of *Binary Decision Diagrams (BDDs)* — which are graph representations of Boolean functions — was introduced by Akers in 1978[Ake78], and efficient methods for manipulating BDDs were developed by Bryant in 1986[Bry86]. BDDs have since attracted the attention of many researchers because of their suitability for representing Boolean functions. We can easily check the equivalence of two functions because a BDD gives a canonical form for a Boolean function. Although a BDD may in the worst case become of exponential size for the number of inputs, its size varies with the kind of functions (unlike a truth table, which always requires 2^n bit of memory). One attractive feature of BDDs is known that many practical functions can be represented by BDDs of feasible sizes.

There have been a number of attempts to improve the BDD technique in terms of execution time and memory space. One of them is the technique of *shared BDDs (SBDDs)*[MIY90], or multi-rooted BDDs, which manages a set of BDDs by joining them into a single graph. This method reduces the amount of memory required and makes it easy to check the equivalence of two BDDs. Another improvement of BDDs is the implementation of *negative edges*, or typed edges[MB88] These are attributed edges such that each edge can have a function of inverting, and they are effective in reducing operation time and the size of the graph.

Boolean function manipulators using BDDs with these improved methods are implemented on workstations and are widely distributed as BDD packages[Bry86, MIY90, MB88, BRB90], that have been utilized in various applications — especially in VLSI CAD systems — such as formal verification[FFK88, MB88, BCMD90], logic synthesis[CMF93, MSB93], and testing[CHJ$^+$90, TIY91].

Despite the advantages of using BDDs to manipulate Boolean functions, there are some problems to be considered in practical applications. One of the problems is variable ordering. Conventional BDDs requires the order of input variables to be fixed, and the size of BDDs greatly depends on the order. It is hard to find the best order that minimizes the size, and the variable ordering algorithm is one of the most important issues in BDD utilization. Another problem occurs because we sometimes manipulate ternary-valued functions containing *don't care* to mask unnecessary information. In such cases, we have to devise a way of representing *don't cares* because the usual BDDs deal only with binary logics. This issue can be generalized to the representation for multi-valued logics or *integer functions*. It is also necessary to efficiently transform BDD representation into other kind of data structures, such as cube sets or Boolean expressions. This is important because in practical applications it is of course necessary to output the result of BDD manipulation.

As our understanding of BDDs has deepened, their range of applications has broadened. Besides having to manipulate Boolean functions, we are often faced with manipulating *sets of combinations* in many problems. One proposal is for multiple fault simulation by representing sets of fault combinations with BDDs[TIY91]. Two others are verification of sequential machines using BDD representation for state sets[BCMD90], and computation of prime implicants using *Meta products*[CMF93], which represent cube sets using BDDs. There is also general method for solving binate covering problems using BDDs[LS90]. By mapping a set of combinations into the Boolean space, we can represent it as a characteristic function using a BDD. This method enables us to manipulate a huge number of combinations implicitly, something that had never been practical before. But because this BDD-based set representation does not completely match the properties of BDDs, the BDDs sometimes grow large because the reduction rules are not effective. There is room to improve the data structure for representing sets of combinations.

1.2 OUTLINE OF THE BOOK

This book discusses techniques related to BDDs and their applications for VLSI CAD systems. The remainder of this book is organized as follows.

In Chapter 2, we start with describing the basic concept of BDDs and Shared BDDs. We then present the algorithms of Boolean function manipulation using BDDs. In implementing BDD manipulators on computers, memory management is an important issue on system performance. We show such implementation techniques that make BDD manipulators applicable to practical problems. As an improvement of BDDs, we propose the use of *attributed edges*, which are the edges attached with several sorts of attributes representing a certain operation. Using these techniques, we developed a BDD subroutine package for Boolean function manipulation. We present the results of some experiments evaluating the applicability of this BDD package to practical problems.

In Chapter 3, we discuss the variable ordering for BDDs. This is important because the size of BDDs greatly depends on the order of the input variables. It is difficult to derive a method that always yields the best order to minimize BDDs, but with some heuristic methods, we can find a fairly good order in many cases. We first consider the general properties of variable ordering for BDDs, and then we propose two heuristic methods of variable ordering: *dynamic weight assignment method* and *minimum-width method*. Our experimental results show that these methods are effective to reduce BDD size in many cases, and are useful for practical applications.

In Chapter 4, we discuss the representation of multi-valued logic functions. In many problems in digital system design, we sometimes use ternary-valued functions containing *don't cares*. The technique of handling *don't care* are basic and important for Boolean function manipulation. We show two methods: *ternary-valued BDDs* and *BDD pairs*, and compare their efficiency. This argument is extended to the functions that deal with integer values. To represent multi-valued logic functions, some variants of BDDs have been devised. We describe these methods and compare them as well as on the ternary-valued functions.

Chapter 5 presents a fast method for generating prime-irredundant cube sets from given BDDs. Prime-irredundant denotes a form such that each cube is a prime implicant and no cube can be eliminated. Our algorithm generates compact cube sets directly from BDDs, in contrast to the conventional cube set reduction algorithms, which commonly manipulate redundant cube sets or truth tables. Experimental results demonstrate that our method is very efficient in terms of time and space.

In Chapter 6, we propose *Zero-suppressed BDDs (ZBDDs)*, which are BDDs based on a new reduction rule. This data structure is adapted to *sets of combinations*, which appear in many combinatorial problems. ZBDDs can manipulate sets of combinations more efficiently than using conventional BDDs. We discuss the properties of ZBDDs, and show their efficiency based on statistical experiments. We then present the basic operators for ZBDDs. Those operators are defined as the operations on sets of combinations, which slightly differ from the Boolean function manipulation based on conventional BDDs.

When describing algorithms or procedures for manipulating BDDs, we usually use Boolean expressions based on switching algebra. Similarly, when considering ZBDDs, we can use *unate cube set* expressions and their algebra. This enables us to easily describe algorithms or procedures for ZBDDs. In this chapter, we present efficient algorithms for computing unate cube set operations, and show some practical applications.

In Chapter 7, an application for VLSI logic synthesis is presented. We propose a fast factorization method for implicit cube set representation based on ZBDDs. In this method, we can quickly factorize multi-level logics from implicit cube sets even for parity functions and full-adders, something that have not been possible with the conventional methods. Experimental results indicate both that our method is much faster than conventional methods and that the differences in speed are more significant for larger-scale problems. Our method greatly accelerates multi-level logic synthesis systems and enlarges the scale of the circuits to which they can be applied.

In Chapter 8, we present another good application of the ZBDD technique, one that manipulates arithmetic polynomial formulas containing higher-degree variables and integer coefficients. This method enables us to represent large-scale polynomials compactly and uniquely and to manipulate them in a practical time. Constructing canonical forms of polynomials immediately leads to equivalence checking of arithmetic expressions. Polynomial calculus is a basic part of mathematics, so it is useful for various problems.

In Chapter 9, we present a helpful tool for research on computer science. Our product, called *BEM-II*, calculates not only binary logic operations but also arithmetic operations on multi-valued logics (such as addition, subtraction, multiplication, division, equality and inequality). Such arithmetic operations provide simple descriptions for various problems. We show the data structure and algorithms for the arithmetic operations. We then describe the specification of BEM-II and some application examples.

In Chapter 10, the conclusion of this book is stated.

2

TECHNIQUES OF BDD MANIPULATION

This chapter introduces the basic concept of BDDs, which are now commonly used for Boolean function representation. We then discuss methods for manipulating BDDs on computers and describe techniques for reducing computation time and memory requirements.

2.1 BINARY DECISION DIAGRAMS

Binary Decision Diagrams (BDDs) are graph representation of Boolean functions, as shown Fig. 2.1(a). The basic concept of BDDs was introduced by Akers[Ake78], and an efficient manipulation method was developed by Bryant[Bry86].

A *BDD* is a directed acyclic graph with two terminal nodes, which we call the *0-terminal node* and *1-terminal node*. Each non-terminal node has an index to identify an input variable of the Boolean function and has two outgoing edges, called the *0-edge* and *1-edge*.

An *Ordered BDD (OBDD)* is a BDD where input variables appear in a fixed order in all the paths of the graph and no variable appears more than once in a path. In this book, we use natural numbers $1, 2, \ldots$ for the indices of the input variables, and every non-terminal node has an index greater than those of its descendant nodes.[1]

A compact OBDD is derived by reducing a binary tree graph, as shown in Fig. 2.1(b). In the binary tree, 0-terminals and 1-terminals represent logic values (0/1), and each

[1] Although this numbering is reversal to that in Bryant's paper[Bry86], it is convenient because the variable index at the root-node gives the number of variables for the function.

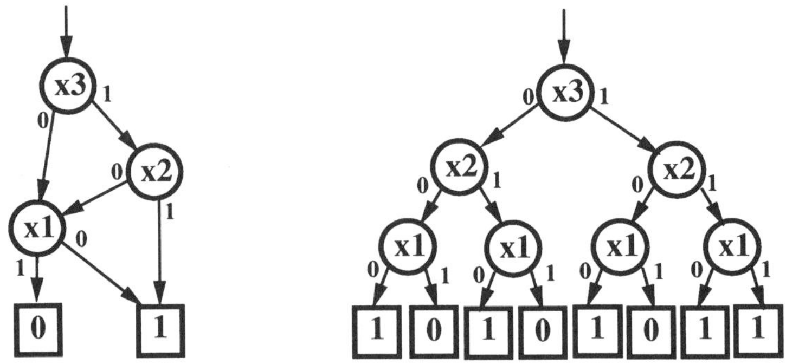

(a) A BDD for $(x3 \cdot x2 \vee \overline{x1})$. (b) A binary decision tree.

Figure 2.1 A BDD and a binary decision tree.

node represents the *Shannon's expansion* of the Boolean function:

$$f = \overline{x_i} \cdot f_0 \vee x_i \cdot f_1,$$

where i is the index of the node and where f_0 and f_1 are the functions of the nodes pointed to by the 0-edge and the 1-edges.

The following reduction rules give a *Reduced Ordered BDD (ROBDD)*:

1. Eliminate all the redundant nodes whose two edges point to the same node. (Fig. 2.2(a))
2. Share all the equivalent subgraphs.(Fig. 2.2(b))

ROBDDs give canonical forms for Boolean functions when the order of variables is fixed. This property is very important for practical applications, since we can easily check the equivalence of two Boolean function simply by checking the isomorphism of their ROBDDs. Most works relating to BDDs are based on the technique of the ROBDDs, so for simplicity in this book, we refer to ROBDDs as BDDs.

Since there are 2^{2^n} kinds of n-input Boolean functions, the representation requires at least 2^n bit of memory in the worst case. It is known that a BDD for an n-input function includes $O(2^n/n)$ nodes in general[LL92]. As each node consumes about $O(n)$ bit (to distinguish the two child nodes from $O(2^n/n)$ nodes), the total storage

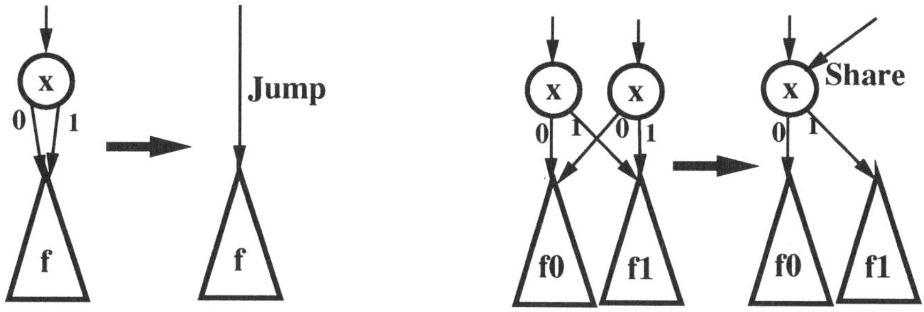

(a) Node elimination. (b) Node sharing.

Figure 2.2 Reduction rules for BDDs.

space exceeds 2^n bit. The size of BDDs, however, varies with the kind of functions, whereas truth tables always require 2^n bit of memory. There is a class of Boolean functions that can be represented by a polynomial size of BDDs, and many practical functions fall into this class[IY90]. This is an attractive feature of BDDs.

A set of BDDs representing multiple functions can be united into a graph that consists of BDDs sharing their subgraphs with each other, as shown in Fig. 2.3. This sharing saves time and space to have duplicate BDDs. When the isomorphic subgraphs are completely shared, two equivalent nodes never coexist. We call such graphs *Shared BDDs (SBDDs)*[MIY90], or *multi-rooted BDDs*.

In the shared BDD environment, the following advantages are obtained:

- Equivalence checking can be performed immediately by just looking at the root nodes.
- We can save the time and space to have duplicate BDDs by only copying a pointer to the root node.

Bryant's original paper[Bry86] presents algorithms for manipulating non-shared (separated) BDDs, but shared BDDs are now widely used and algorithms more concise than original ones are available. In the rest of this book, we assume the shared BDD environment.

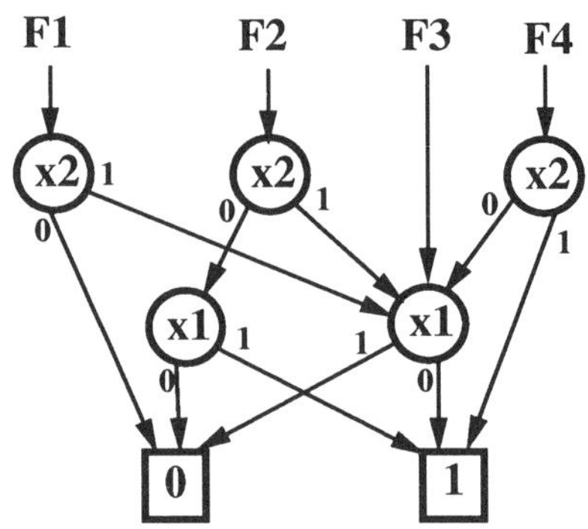

$(F1 = x2 \cdot \overline{x1},\ F2 = x2 \oplus x1,\ F3 = \overline{x1},\ F4 = x2 \vee \overline{x1}.\ )$

Figure 2.3 A shared BDD.

(Address)	(Index)	(0-edge)	(1-edge)	
N_0	-	-	-	$\longleftarrow 0$
N_1	-	-	-	$\longleftarrow 1$
N_2	x_1	N_0	N_1	
N_3	x_1	N_1	N_0	$\longleftarrow F_3(=\overline{x_1})$
N_4	x_2	N_0	N_3	$\longleftarrow F_1(=x_2 \cdot \overline{x_1})$
N_5	x_2	N_2	N_3	$\longleftarrow F_2(=x_2 \oplus x_1)$
N_6	x_2	N_3	N_1	$\longleftarrow F_4(=x_2 \vee \overline{x_1})$

Figure 2.4 BDD representation using a table.

In a typical implementation of a BDD manipulator, all the nodes are stored in a *node table* on the main memory of the computer. Figure 2.4 shows, as an example, a node table representing the BDDs shown in Fig. 2.3. Each node has three basic attributes: an index of the input variable and two pointers of the 0- and 1-edges. Some additional pointers and counters for maintaining the node table are attached to the node data. The 0- and 1-terminal nodes are at first allocated in the table as special nodes, and the other (non-terminal) nodes are gradually generated as the results of logic operations. By referring to the address of a node, we can immediately determine whether or not the node is a terminal.

In the shared BDD environment, all isomorphic subgraphs are shared. That is, two equivalent nodes should never coexist. To assure this, before creating a new node we always check the reduction rules shown in Fig. 2.2. If the 0- and 1-edges have the same destination or if an equivalent node already exists, we do not create a new node but simply copy the pointer to the existing node. We use a hash table which indices all the nodes, so that the equivalent node can be checked in a constant time if the hash table acts successfully. The effectiveness of the hash table is important since it is frequently referred to in the BDD manipulation. In this technique, the uniqueness of the nodes is maintained, and every Boolean function can be identified by a 1-word address of the root node.

In a typical implementation, the BDD manipulator requires 20 to 30 Bytes of memory for each node. Today's workstations with more than 100 MByte of memory enable us to generate BDDs containing millions of nodes, but BDDs still grow beyond the memory capacity in some applications.

2.2 LOGIC OPERATIONS

In many cases, BDDs are generated as the results of logic operations. Figure 2.5 shows an example for $(x_1 \cdot x_2) \vee x_3$. First, trivial BDDs for x_1, x_2, x_3 are created. Then by applying the AND operation between x_1 and x_2, the BDD for $(x_1 \cdot x_2)$ is generated. The final BDD for the entire expression is obtained as the result of the OR operation between $(x_1 \cdot x_2)$ and x_3.

The basic logic operations are summarized as follows:

- Creating a BDD for a single variable function x_i.
- Generating $\overline{f}$ for a given BDD f.
- Generating a BDD as the result of a binary operation $(f \circ g)$, that includes $(f \cdot g), (f \vee g)$ and $(f \oplus g)$.
- Generating a BDD for $f_{(x_i=0)}, f_{(x_i=1)}$.
- Equivalence checking of two BDDs.
- Finding the input assignments to satisfy $f = 1$.

In this section we show the algorithms of logic operations on BDDs.

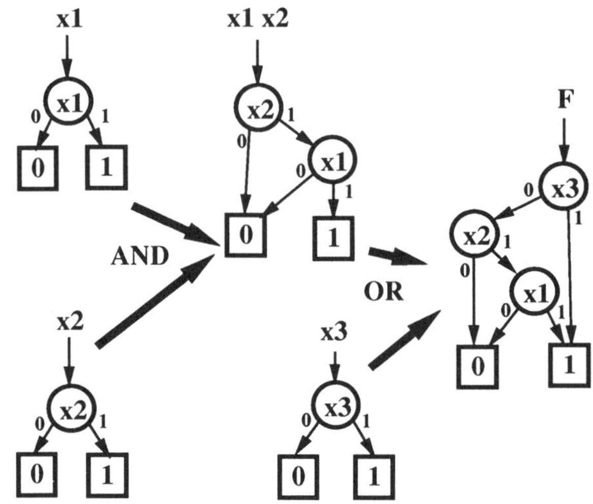

Figure 2.5 Generation of BDDs for $F = (x1 \cdot x2 \vee x3)$.

2.2.1 Binary Operations

The binary logic operation $(f \circ g)$ is the most important part of the techniques of BDD manipulation. Its computation is based on the following expansion:

$$f \circ g = \overline{v} \cdot (f_{(v=0)} \circ g_{(v=0)}) \vee v \cdot (f_{(v=1)} \circ g_{(v=1)}),$$

where v is the highest ordered variable in f and g. This expansion creates a new node with the variable v having two subgraphs generated by suboperations $(f_{(v=0)} \circ g_{(v=0)})$ and $(f_{(v=1)} \circ g_{(v=1)})$. This formula means that the operation can be expanded to two suboperations $(f_{(v=0)} \circ g_{(v=0)}))$ and $(f_{(v=1)} \circ g_{(v=1)})$ with respect to an input variable v. Repeating this expansion recursively for all the input variables, eventually a trivial operation appears, and the result is obtained. (For instance, $f \cdot 1 = f$, $f \oplus f = 0$, etc.) As mentioned in the previous section, we check the reduction rules before creating a new node.

The algorithm of computing h $(= f \circ g)$ is summarized as follows. (Here $f.top$ denotes the input variable of the root node of f, and f_0 and f_1 are the BDDs pointed to by the 0-edge and the1-edge from the root node.)

1. When f or g is a constant or when $f = g$:
   return a result according to the kind of the operation.
   (Example) $f \cdot 0 = 0,\ \ f \vee f = f,\ \ f \oplus 1 = \overline{f}$

2. If $f.top$ and $g.top$ are identical:
   $h_0 \leftarrow f_0 \circ g_0;\quad h_1 \leftarrow f_1 \circ g_1;$
   if $(h_0 = h_1)\ h \leftarrow h_0;$
   else $h \leftarrow Node(f.top,\ h_0,\ h_1);$

3. If $f.top$ is greater than $g.top$:
   $h_0 \leftarrow f_0 \circ g;\quad h_1 \leftarrow f_1 \circ g;$
   if $(h_0 = h_1)\ h \leftarrow h_0;$
   else $h \leftarrow Node(f.top,\ h_0,\ h_1);$

4. If $f.top$ is less than $g.top$:
   (Compute similarly to 3. by exchanging f and g.)

As mentioned in previous section, we check the hash table before creating a new node to avoid duplication of the node.

Figure 2.6(a) shows an example. When we perform the operation between the nodes (1) and (5), the procedure is broken down into the binary tree, shown in Fig. 2.6(b). We usually compute this tree in a *depth-first manner*. It seems that this algorithm always takes a time exponential for the number of inputs, since it traverses the binary trees. However, some of these trees are redundant; for instance, the operations of (3)-(4) and (4)-(7) appear more than once. We can accelerate the procedure by using a hash-based cache that memorizes the results of recent operations. By referring to the cache before every recursive call, we can avoid duplicate executions for equivalent suboperations. This technique enables the binary logic operations to be executed in a time almost proportional to the size of the BDDs.

The cache size is important for the performance of the BDD manipulation. If the cache size is insufficient, the execution time grows rapidly. Usually we fix the cache size empirically, and in many cases it is several times greater or smaller than the number of the nodes.

2.2.2 Negation

A BDD for $\overline{f}$, which is the complement of f, has a form similar to that of the BDD for f: just the 0-terminal and the 1-terminal are exchanged. Complementary BDDs

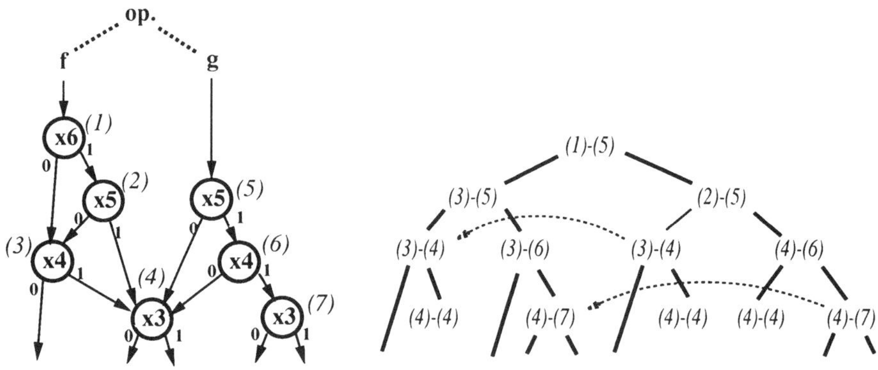

(a) An example. (b) Structure of procedure calls.

Figure 2.6 Procedure of binary operation.

contain the same number of nodes, in contrast to the cube set representation, which sometimes suffers a great increase of the data size.

By using binary operation $(f \oplus 1)$, we can compute $\overline{f}$ in a time linear to the size of the BDDs. However, the operation is improved to a constant time by using the *negative edge*. The negative edges are a kind of attributed edges, discussed in Section 2.4. This technique is now commonly used in many implementations.

2.2.3 Restriction (Cofactoring)

After generating a BDD for f, we sometimes need to compute $f_{(v=0)}$ or $f_{(v=1)}$, such that an input variable is fixed to 0 or 1. This operation is called *restriction*, or *cofactoring*. If v is the highest ordered variable in f, a BDD pointed to by the 0- or the 1-edge of the root node is just returned. Otherwise we have to expand the BDDs until x become the highest ordered variable, and then re-combine them into one BDD. This procedure can be executed efficiently by using the cache technique as the binary operations. The computation time is proportional to the number of nodes that have an index greater than v.

2.2.4 Search for Satisfiable Assignment

When a BDD for f has been generated, it is easy to find an assignment of input variables to satisfy the function $f = 1$. If there is a path from the root node to the 1-terminal node (which we call *1-path*), the assignment for variables to activate the 1-path is a solution for $f = 1$. An excellent property of BDDs is that every non-terminal node is included in at least one 1-path. (This is obvious since if there is no 1-path, the node should be reduced into the 0-terminal node.) By traversing the BDD from the root node, we can easily find a 1-path in a time proportional to the number of the input variables, independently of the number of nodes.

In general, there are many solutions to satisfy a function. Under the definition of the costs to assign "1" to respective input variables, we can search for an assignment that minimizes the total cost[LS90], which is defined as $\sum_{i=1}^{n}(C_i \times x_i)$, where C_i is a non-negative cost for input variable $x_i (\in \{0, 1\})$. Many NP-complete problems can be described in the above format.

Searching for the minimum-cost 1-path can be implemented by a backtrack search of the BDD. It appears to take an exponential time, but we can avoid duplicate tracking for shared subgraphs in the BDD by storing the minimum cost for the subgraph and referring to it at the second visit. This technique eliminates the need to visit each node more than once, so we can find the minimum cost 1-path in a time proportional to the number of nodes in the BDD.

With this method, we can immediately solve the problem if the BDD for the constraint function can be generated in the main memory of the computer. There are many practical examples where the BDD becomes compact. Of course, it is still an NP problem so in the worst case the BDD requires an exponential number of nodes and overflows the memory.

We can also efficiently count the number of the solutions to satisfy $f = 1$. On the root node of f, the number of solutions is computed as the sum of the solutions of the two subfunctions f_0 and f_1. By using the cache technique to save the result on each node, we can compute the number of the solutions in a time proportional to the number of nodes in the BDD.

In a similar way, we can compute the truth table density for a given Boolean function represented by a BDD. The truth table density is the rate of 1's in the truth table, and it indicates the probability of satisfying $f = 1$ for an arbitrary assignment to the input variables. Using BDDs, we can compute it as an average of the density for the two subfunctions on each node.

2.3 MEMORY MANAGEMENT

When generating BDDs for Boolean expressions, many intermediate BDDs are temporarily generated. It is important for the memory efficiency to delete such already used BDDs. If the used BDD shares subgraphs with other BDDs, we cannot delete the subgraphs. In order to determine the necessity of the nodes, a *reference counter* that shows the number of incoming edges is attached to each node. When a BDD becomes unnecessary, we decrease the reference counter of the root node, and if it becomes zero, the node can be eliminated. When a node is really deleted, we recursively execute this procedure on its descendant nodes. On the other hand, in the case of copying an edge to a BDD, the reference counter of the root node is increased.

Deletion of the used nodes saves memory space, but there may be a time cost because we might eliminate nodes that will later become necessary again. Besides, deletion of the nodes breaks the consistency of the cache for memorizing recent operation, so we have to recover (or clear) the cache. To avoid these problems, even when the reference counter becomes zero, we do not eliminate the nodes until the memory becomes full. Then they are deleted all at once as a *garbage collection*.

The BDD manipulator is based on the hash table technique, so we have to allocate a fixed size of hash table when initializing the program. At that time, it is difficult to estimate the final size of BDDs to be generated, but it is inefficient to allocate too much memory because other application programs cannot use the space. In our implementation, some small amount of space is allocated initially, and if the table becomes full during BDD manipulation, the table is re-allocated two or four times larger (unless the memory overflows). When the table size reaches the limit of extension, the garbage collection process is invoked.

2.4 ATTRIBUTED EDGES

The *attributed edge* is the technique to reduce the computation time and memory required for BDDs by using edges with an attribute representing a certain operation. Several kinds of attributed edges have been proposed[MIY90]. In this section, we describe three kinds of attributed edges: *negative edges, input inverters*, and *variable shifters*. The negative edge is particularly effective and now widely implemented in BDD manipulators.

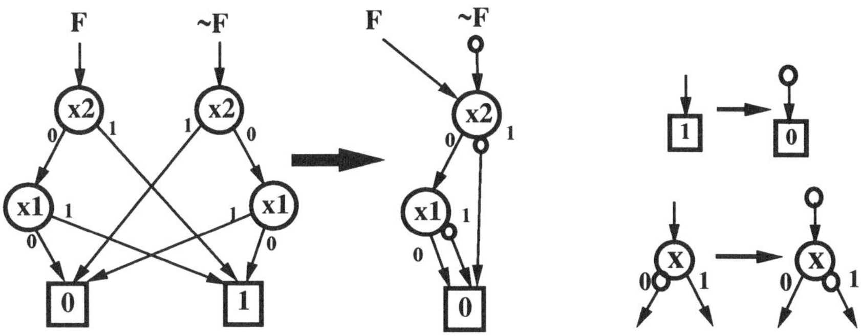

(a) Effect of negative edges. (b) Rule of usage.

Figure 2.7 Negative edges.

2.4.1 Negative edges

The negative edge has an attribute which indicates that the function of the subgraph pointed by the edge should be complemented, as shown in Fig. 2.7(a). This idea was introduced as Akers's *inverter*[Ake78] and Madre and Billon's *typed edge*[MB88]. The use of negative edges results in some remarkable improvements:

- The BDD size is reduced by as much as one half.
- Negation can be performed in a constant time.
- Logic operations can be accelerated by applying rules such as $f \cdot \overline{f} = 0$, $f \vee \overline{f} = 1$, and $f \oplus \overline{f} = 1$.
- Using the quick negation, we can transform operations such as *OR, NOR*, and *NAND* into *AND* by applying *De Morgan's theorems*, thus raising the hit rate of the cache of recent operations.

The use of negative edges may break the uniqueness of BDDs. To preserve the uniqueness, we have to put two constraints:

1. Using 0-terminal node only.
2. Not using a negative edge at the 0-edges on each node.

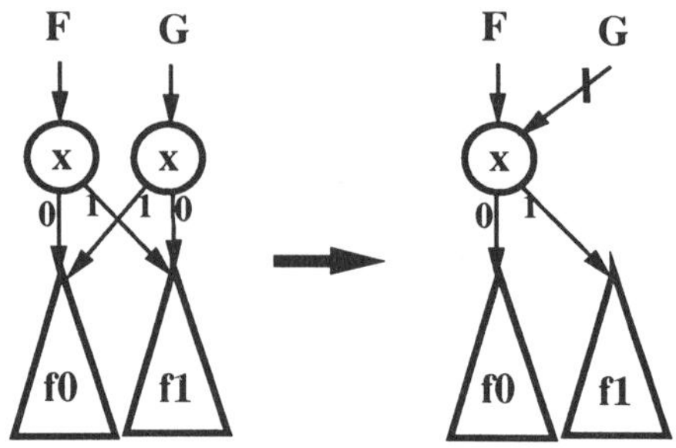

Figure 2.8 Input inverters.

If necessary, the negative edges can be carried over as shown in Fig. 2.7(b). These constraints are basically same as those in Madre and Billon's work[MB88].

2.4.2 Input Inverters

We propose another attribute indicating that the 0- and 1-edges at the next node should be exchanged (Fig. 2.8). Since it is regarded as complementing an input variable of the node, we call it an *input inverter*. Using input inverters, we can reduce the size of BDDs by as much as half. There are cases where input inverters are very effective while negative edges are not (Fig. 2.9).

Since abuse of input inverters can also breaks the uniqueness of BDDs, we also place a constraint on their use. We use input inverters so that the two nodes f_0 and f_1 pointed to by the 0- and 1-edges satisfy the constraint: $(f_0 < f_1)$, where '$<$' represents an arbitrary total ordering of all the nodes. In our implementation, each edge identifies the destination with the address of the node table, so we define the order as the value of the address. Under this constraint, the BDD forms may vary with the addressing manner, but the uniqueness is assured in the shared BDD environment.

2.4.3 Variable Shifters

When there are the two subgraphs that are isomorphic except for a difference of the index numbers of their input variables, we want to combine them into one subgraph by storing the difference of the indices. We therefore propose *variable shifters*, which

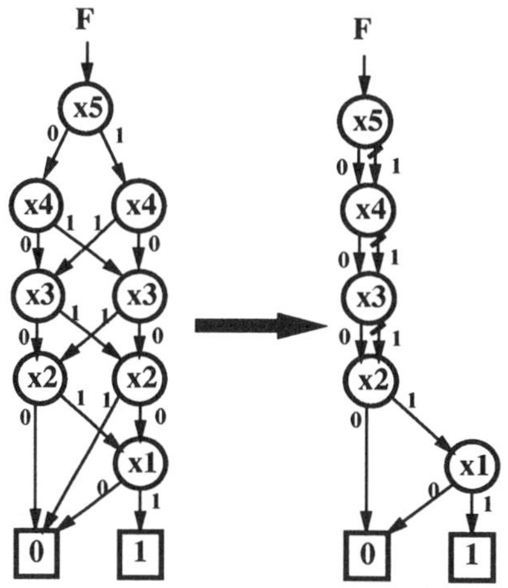

Figure 2.9 An example where input inverters are effective.

indicate that a number should be added to the indices of all its descendant nodes. We do not record information of variable index on each node because the variable shifter gives information about the distance between the pair of nodes. We place the following rules to use variable shifters:

1. On the edge pointing to a terminal node, do not use a variable shifter.
2. On the edge pointing to the root node of a BDD, the variable shifter indicates the absolute index number of the node.
3. Otherwise, a variable shifter indicates the difference between the index numbers of the start node and the end node of the edge.

As shown in Fig. 2.10, for example, the graphs representing $(x_1 \vee (x_2 \cdot x_3))$, $(x_2 \vee (x_3 \cdot x_4)), \ldots, (x_k \vee (x_{k+1} \cdot x_{k+2}))$ can be joined into the same graph.

Using variable shifters, we can reduce the size of BDDs — especially when manipulating a number of regular functions, such as arithmetic systems. Another advantage of using variable shifters is that we can raise the hit rate of the cache of operations by applying the rule:

$$(f \circ g = h) \Longleftrightarrow (f^{(k)} \circ g^{(k)} = h^{(k)}),$$

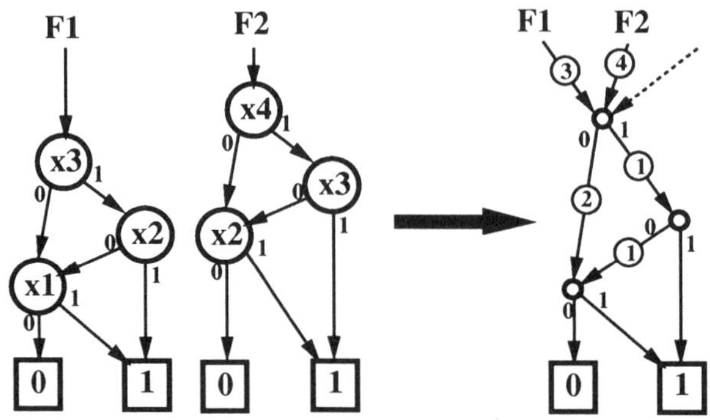

Figure 2.10 Variable shifters.

where $f^{(k)}$ is a function whose indices are shifted by k from f and where $\circ$ is a binary logic operation such as *AND*, *OR*, and *EXOR*.

2.4.4 General Consideration

Here we show a general method to define an attributed edge. Let S be the set of all the Boolean functions to be represented:

1. Divide S into the two subsets S_0 and S_1
2. Define a mapping $\mathcal{F} : (S \rightarrow S)$, such that for any $f \in S_1$ there is a unique $f_0 \in S_0$ to satisfy $f = \mathcal{F}(f_0)$.

$\mathcal{F}$ represents the operation of the attributed edge. By applying $\mathcal{F}$, every subgraph for a function in S_1 can be represented by one in S_0, and we use the subgraphs only in S_0. This promotes BDD reduction up to a half. Notice that the reversal mapping from S_0 to S_1 breaks the uniqueness of BDD representation, so that we have to keep the rule on the location of using attributed edges.

The above argument explains both negative edges and input inverters. Variable shifters can also be explained if we extend the argument having more than two partitions, as follows:

1. Partition S into a number of subsets $S_0, S_1, \ldots S_n$.
2. For any $k \geq 1$, define a mapping $\mathcal{F}_k : (S \rightarrow S)$, such that for any $f_k \in S_k$ there is a unique $f_0 \in S_0$ to satisfy $f_k = \mathcal{F}_k(f_0)$.

2.5 IMPLEMENTATION AND EXPERIMENTS

In this section, we describe the implementation of a BDD package and present the results of experiments evaluating the performance of the BDD package and the effect of attributed edges.

2.5.1 BDD Package

We implemented an BDD program package on a Sun3/60 workstation (24Mbyte, SunOS 4.0). The program consists of about 800 lines of C code, and this package supports the following basic and the essential operations of Boolean functions:

- Giving the trivial functions, such as 1 (tautology), 0 (inconsistency), and x_k for a given index k.
- Generating BDDs by applying logic operations such as *NOT*, *AND*, *OR*, *EXOR*, and *restriction*.
- Equivalence or implication checking of two functions.
- Finding an assignment of inputs to satisfy a function.
- Copying and deleting BDDs.

These operations are modularized as functions of C language. By using them in combination, we can utilize the package for various applications without considering the detailed structure of the program. In this package we implemented attributed edges (negative edges, input inverters, and variable shifters). The storage requirement of this package is about 22 Bytes per node, and we can manage up to $700,000$ nodes on our machine.

Table 2.1 Experimental results.

Circuit	Circuit size			BDD	Time
	Inputs	Outputs	Nets	nodes	(sec)
sel8	12	2	29	78	0.3
enc8	9	4	31	56	0.3
add8	18	9	65	119	0.4
add16	33	17	129	239	0.7
mult4	8	8	97	524	0.5
mult8	16	16	418	66161	24.8
c432	36	7	203	131299	55.5
c499	41	32	275	69217	22.9
c880	60	26	464	54019	17.5
c1355	41	32	619	212196	89.9
c1908	33	25	938	72537	33.0
c5315	178	123	2608	60346	31.3

2.5.2 Experimental Results

To evaluate the efficiency of the program, we generated BDDs from combinational circuits. Notice that in this experiment the BDDs represent the set of the functions of not only all primary outputs but all the internal nets. To count the number of nodes exactly, at last we execute garbage collection, that is unnecessary in practical use. We used the heuristic variable ordering: *Dynamic weight assignment method*, described in Chapter 3.

The results are listed in Table 2.1. The circuit "sel8" is an 8-bit data selector, and "enc 8" is an 8-bit encoder. The circuits "add8" and "add16" are 8-bit and 16-bit adders, and "mult4" and "mult8" are 4-bit and 8-bit multipliers. The rests are selected from the benchmarks in ISCAS'85[BF85]. The column "BDD nodes" lists the number of the nodes in the set of BDDs, and the total time for loading the circuit data, ordering the input variables, and generating the BDDs is listed in column "Time (sec)."

The results show that we can represent the functions of these practical circuits quickly and compactly. It took less than a minute to represent circuits with dozens of inputs and hundreds of nets. We can see that the CPU time is almost proportional to the number of nodes.

Table 2.2 Effect of attributed edges.

Circuit	(A) BDD nodes	(A) Time (sec)	(B) BDD nodes	(B) Time (sec)	(C) BDD nodes	(C) Time (sec)	(D) BDD nodes	(D) Time (sec)
sel8	78	0.3	51	0.3	51	0.4	40	0.3
enc8	56	0.3	48	0.3	48	0.3	33	0.3
add8	119	0.4	81	0.4	81	0.4	49	0.4
add16	239	0.7	161	0.6	161	0.7	97	0.6
mult4	524	0.5	417	0.5	400	0.4	330	0.5
mult8	66161	24.8	52750	19.1	50504	19.8	46594	18.3
c432	131299	55.5	104066	36.5	103998	36.8	89338	34.1
c499	69217	22.9	65671	21.3	36986	21.8	36862	21.5
c880	54019	17.5	31378	10.8	30903	11.1	30548	11.5
c1355	212196	89.9	208324	49.3	119465	52.8	119201	51.4
c1908	72537	33.0	60850	21.6	39533	22.3	39373	22.5
c5315	60346	31.3	48353	29.2	41542	28.6	40306	29.8

(A): Using nothing, (B): (A)+ output inverters,
(C): (B)+ input inverters, (D): (C) + variable shifters

To evaluate the effect of the attributed edges, we conducted the similar experiments by incrementally applying the three kinds of attributed edges. The results are shown in Table 2.2. Column (A) lists the results of the experiments using original BDDs without any kind of attributed edges, and column (B) lists the results obtained when using only negative edges. Comparing the results, we can see that the use of negative edges reduces the number of nodes by as much as 40% and speed up the computation markedly.

Column (C) lists the results obtained when using both input inverters and negative edges. The number of nodes is reduced owing to the use of input inverters, but there are no remarkable differences in CPU time. Input inverters were especially effective for the circuits "c499," "c1355," and "c1908," for which we can see up to 45% reduction in size.

And as we can see from the results listed in column (D), the additional use of the variable shifters reduced graph size still more without producing remarkable differences in the CPU time. Variable shifters are effective especially for the circuits with regular structures, such as arithmetic logics, and we can also see some effectiveness for other circuits.

These experimental results show that the combination of the three attributed edges is very effective in many cases, though none of them are themselves effective for all types of circuits.

2.6 CONCLUSION

In this chapter, we presented basic algorithms of Boolean function manipulation using BDDs, and we then described implementation techniques to make BDD manipulators applicable to practical problems. As an improvement of BDDs, we proposed the use of *attributed edges*, which are the edges attached with several kinds of attributes representing a certain operation. Particularly, the negative edges are now widely used because of their remarkable advantages. In the rest of this book, we show experimental results of BDD manipulation with negative edges.

Using these techniques, we implemented a BDD subroutine package. Experimental results show that we can efficiently manipulate very large-scale BDDs containing more than million of nodes. Such BDD manipulators have been developed and improved in many laboratories[Bry86, MIY90, MB88, BRB90], and some program packages are opened as the public domain software. A number of works using those BDD packages are in progress. BDDs and their improvements will become the key techniques for solving various problems in computer science.

3

VARIABLE ORDERING FOR BDDS

BDDs give canonical forms of Boolean functions provided that the order of input variables is fixed, but the BDD for a given function can have many different forms depending on the permutation of the variables — and sometimes the size of BDDs greatly varies with the order. The size of BDDs determines not only the memory requirement but also the amount of execution time for their manipulation. The variable ordering algorithm is thus one of the most important issues in the application of BDDs.

The effect of variable ordering depends on the kind of function to be handled. There are very sensitive functions whose BDDs vary extremely (exponentially to the number of inputs) by only reversing the order. Such functions often appear in practical digital system designs. On the other hand, there are some kinds of functions for which the variable ordering is ineffective. For The symmetric functions, for example, obviously have the same form for any variable order. It is known that the functions of multipliers[Bry91] cannot be represented by a polynomial-sized BDD in any order.

There are some works on the variable ordering. Concerning the method to find exactly the best order, Friedman et al. presented an algorithm[FS87] of $O(n^2 3^n)$ time based on the *dynamic programming*, where n is the number of inputs. For functions with many inputs, it is still difficult to find the best order in a practical time, although this algorithm has been improved to the point that the best order for some functions with 17 inputs can be found[ISY91].

From the practical viewpoint, heuristic have been studied extensively. Malik et al.[MWBSV88] and Fujita et al.[FFK88] reported heuristic methods based on the topological information of logic circuits. Butler et al. [BRKM91] uses a testability measure for the heuristics, which reflect not only topological but logical information of the circuit. These methods can find a (may be) good order before generating BDDs.

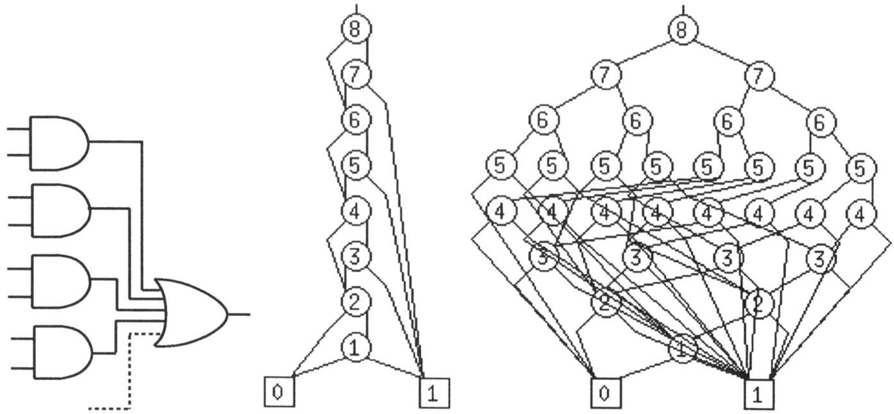

(a) Circuit. (b) In the best order. (c) In the worst order.

Figure 3.1 BDDs for a 2-level AND-OR circuit.

They are applied to the practical benchmark circuits and in many cases compute a good order.

Fujita et al.[FMK91] showed another approach that improves the order for the given BDD by repeating the exchange of the variables. It can give results better than the initial BDDs, but it is sometimes trapped in a local optimum.

In this chapter, we discuss the properties of the variable ordering for BDDs and show two heuristic methods we have developed for variable ordering.

3.1 PROPERTIES OF THE VARIABLE ORDERING

The variable ordering of BDDs has the following empirical properties:

1. *(Local computability)*
   The group of the inputs with local computability should be near in the order. Namely, inputs that are closely related should be kept near to each other. Consider, for example, the BDD representing the function of the *AND-OR* 2-level logic circuit with $2n$ inputs shown in Fig. 3.1(a). It takes $2n$ nodes under the

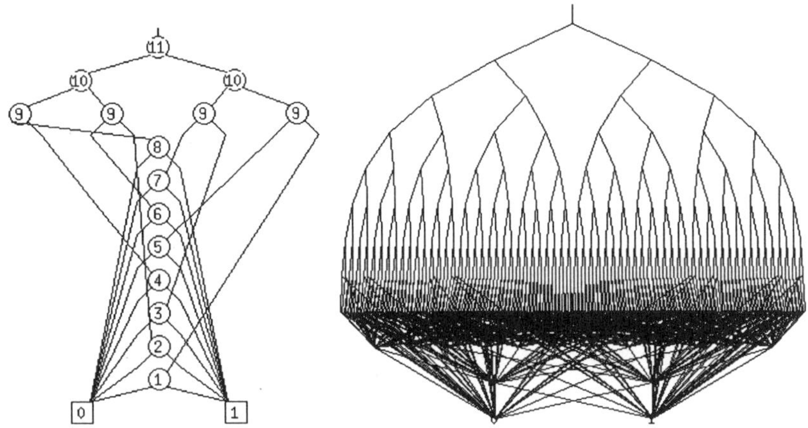

(a) In the best order. (b) In the worst order.

Figure 3.2 BDDs for an 8-bit data selector.

ordering $x_1 \cdot x_2 \vee x_3 \cdot x_4 \vee \ldots \vee x_{2n-1} \cdot x_{2n}$ (Fig. 3.1(b)), but under the ordering $x_1 \cdot x_{n+1} \vee x_2 \cdot x_{n+2} \vee \ldots \vee x_n \cdot x_{2n}$ it takes $(2 \cdot 2^n - 2)$ nodes (Fig. 3.1(c)).

2. *(Power to control the output)*
   The inputs that greatly affect the function should be located at higher positions in the order (nearer positions to the root node). Consider the BDD representing function of an 8-bit data selector with three control inputs and eight data inputs. When the control inputs are high in the order, the BDD size is linear (Fig. 3.2(a)), whereas it becomes an exponential number of nodes using the reversal order (Fig. 3.2(b)).

If we could find a variable order that satisfies those two properties, the BDDs would be compact. However, the two properties are mixed ambiguously, and sometimes they require the conflicting orders with each other. Another problem is that when multiple functions are represented together, those functions may require different orders. It is difficult to find a point of compromise. For large-scale functions, automatic methods giving an appropriate solution are desired.

There are two approaches to the heuristic methods of variable ordering:

- To find an appropriate order before generating BDDs by using the logic circuit information that is the source of the Boolean function to be represented.
- To reduce BDD size by permuting the variables of a given BDD started with an initial variable order.

The former approach refers only to the circuit information and does not use the detailed logical data, to carry out the ordering in a short time. Although it sometimes gives a poor result depending on the structure of the circuits, it is still one of the most effective ways to deal with large-scale problems. The latter approach, on the other hand, can find a fairly good order by fully using the logical information of the BDDs. It is useful when the former method is not available or ineffective. A drawback of this approach is that it cannot start if we fail to make an initial BDD of reasonable size.

We first propose *Dynamic Weight Assignment (DWA) method*, which belongs the former approach, and show their experimental results. We then present *Minimum-width method*, which is an reordering method we developed. The two methods can be used in combination.

3.2 DYNAMIC WEIGHT ASSIGNMENT METHOD

When we utilize BDD techniques for digital system design, we first generate BDDs representing the functions for the given logic circuits. If the initial BDDs are intractably large, we cannot perform any BDD operation. Therefore, it is important for practical use to find a good order before generating BDDs.

Considering the properties of the variable ordering of BDDs, we have developed the *Dynamic Weight Assignment () method*[MIY90], in which the order is computed from the topological information of a given combinational circuit.

3.2.1 Algorithm

We first assign a weight 1.0 to one of the primary outputs of the circuit, and then that weight is propagated toward the primary inputs in the following manner:

1. At each gate, the weight on the output is divided equally and distributed to the inputs.

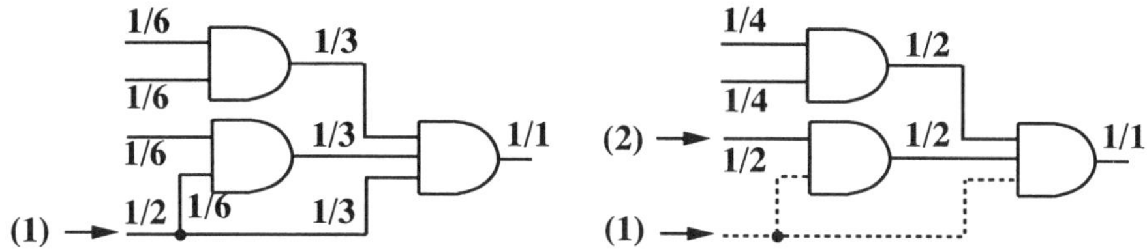

(a) First assignment.

(b) Second assignment.

Figure 3.3 Dynamic weight assignment method.

2. At each fan-out point, the weights of the fan-out branches are accumulated into the fan-out stem.

After this propagation, we give the highest order to the primary input with the largest weight. Since the weights reflect the contribution to the primary output in a topological sense, primary inputs with a large weight are expected to have a large influence to the output function.

Next, after choosing the highest input, we delete the part of the circuit that can be reached only from the primary input already chosen, and we re-assign the weights from the beginning to choose the next primary input. By repeating this assignment and deletion, we obtain the order of the input variables. An example of this procedure is illustrated in Fig. 3.3(a) and (b). When we delete the subcircuit, the largest weight in the prior assignment is distributed to the neighboring inputs in the re-assignment, and their new weights are increased. Thereby, the neighboring inputs tend to be close to the prior ones in the order.

When the circuit has multiple outputs, we have to choose one in order to start the weight assignment. We start from the output with the largest logical depth from the primary inputs. If we have not yet ordered all the inputs after the ordering, the output with the next largest depth is selected to order the rest of the inputs.

This algorithm gives a good order in many cases, and the time complexity of this method is $O(m \cdot n)$, where m is the number of the gates and n is the number of the primary inputs. This time complexity is enough tolerable for many practical uses.

Table 3.1 Effect of variable ordering.

Circuit	(A) BDD nodes	(A) Time (sec)	(B) BDD nodes	(B) Time (sec)	(C) BDD nodes	(C) Time (sec)	(D) BDD nodes	(D) Time (sec)
sel8	23	0.1	16	0.1	510	0.1	88	0.1
enc8	28	0.1	28	0.1	23	0.1	29	0.1
add8	41	0.2	83	0.1	222	0.2	885	0.2
add16	81	0.3	171	0.2	830	0.4	19224	1.4
mult4	150	0.1	139	0.1	145	0.1	153	0.2
mult8	10766	2.7	9257	2.4	9083	2.2	15526	4.1
c432	27302	6.1	3987	1.2	1732	0.7	(>500k)	-
c499	52369	7.2	115654	11.6	45921	7.7	(>500k)	-
c880	23364	2.1	(>500k)	-	(>500k)	-	(>500k)	-
c1355	52369	8.1	115654	15.4	45921	8.4	(>500k)	-
c1908	17129	5.1	23258	4.8	36006	9.7	77989	23.5
c5315	31229	9.4	57584	8.1	(>500k)	-	(>500k)	-

(A): Using dynamic weight assignment method, (B): In the original order,
(C): In the original order (reverse), (D): In a random order

3.2.2 Experimental Results

We implemented the DWA method using our BDD package on SPARC station 2 (32Mbyte, SunOS 4.1.3), and conducted the experiments to evaluate the effect of the ordering method. The results are listed in Table 3.1. The circuits are the same ones used in Table 2.2. The column headed "BDD nodes" gives the number of the nodes of BDDs for the output functions, not including internal nets. We used negative edges. "Time (sec)" shows the total time of loading the circuit data, ordering the input variables, and generating the BDDs. The columns (B),(C) and (D) shows the results of the experiments without the heuristic method of variable ordering. For column (B) we use the original order of the circuit data, for column (C) the order is original but the reverse of that for (B), and for column (D) the order is random.

The ordering method is very effective except in a few cases which are insensitive to the order. The random ordering is quite impractical. The use of the original order of the circuit data sometimes gives a good result, but it is a passive way and it cannot always be good. We can conclude that our ordering method is useful and essential for many practical applications.

3.3 MINIMUM-WIDTH METHOD

In this section, we describe a heuristic method based on reordering the variables for a given BDD with an initial order[Min92b]. In the following, n denotes the number of the input variables.

Several ordering methods based on the reordering approach have been proposed. Fujita et al.[FMK91] presented an incremental algorithm based on the exchange of a pair of variables (x_i, x_{i+1}), and Ishiura et al.[ISY91] presented a *simulated annealing method* with the random exchange of two variables. A drawback of these incremental search methods, though, is that their performance depends very much on the initial order. If it is far from the best, many exchanges are needed. This takes longer time and increases the risk of being trapped in a bad local minimum solution.

We therefore propose another strategy. We first choose one variable based on a certain cost function, fix it at the highest position (x_n), and then choose another variable and fix it at the second highest position (x_{n-1}). In this manner, all the variables are chosen one by one, and they are fixed from the highest to the lowest. This algorithm has no backtracking and is robust to the bad initial order. In our method, we define the width of BDDs, as a cost function.

3.3.1 The Width of BDDs

When choosing $x_k (1 \leq k \leq n)$, the variables with indices higher than k have already been fixed and the form of the higher part of the graph never varies. Thus, the choice of x_k affects only the part of the graph lower than x_k. The goal in each step is to choose the x_k that minimizes the size of the lower part of the graph. The cost function should give a good estimation of the minimum size of the graph for each choice of x_k, and should be computable within a feasible time. As a cost function to satisfy those requirements, we define the width of BDDs here.

Definition 3.1 *The* width of a BDD at height k, *denoted* $width_k$, *is the number of edges crossing the section of the graph between* x_k *and* x_{k+1}, *where the edges pointing to the same node are counted as one. The width between* x_1 *and the bottom of graph is denoted* $width_0$. □

An example is shown in Fig. 3.4. the $width_k$ is sometimes greater than the number of the nodes of x_k because the width also includes edges that skip the nodes of x_k.

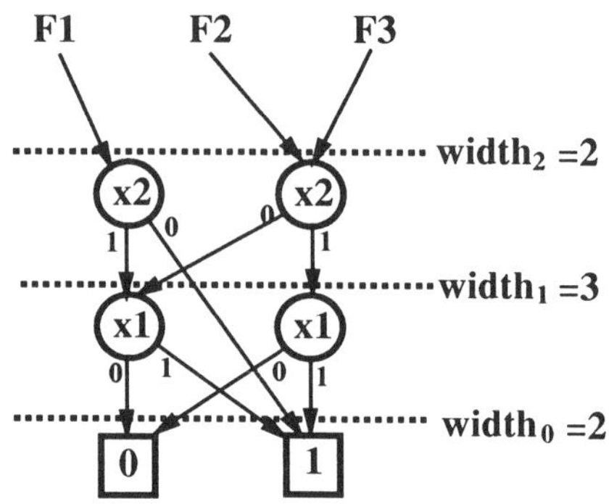

Figure 3.4 Width of a BDD.

We present the following theorem on the width of BDDs.

Theorem 3.1 *The $width_k$ is constant for any permutation of $\{x_1, x_2, \ldots, x_k\}$ and any permutation of $\{x_{k+1}, x_{k+2}, \ldots, x_n\}$.*

(Proof) $width_k$ represents the total number of subfunctions obtained by assigning all the combinations of Boolean values $\{0, 1\}^{n-k}$ to the variables $\{x_{k+1}, x_{k+2}, \ldots, x_n\}$. Because the total number of subfunctions is independent of the order of the assignment, the $width_k$ is constant for any permutation of $\{x_{k+1}, x_{k+2}, \ldots, x_n\}$.

All the subfunctions obtained by assigning all the combinations of Boolean value $\{0, 1\}^{n-k}$ to the variables $\{x_{k+1}, x_{k+2}, \ldots, x_n\}$ are uniquely represented by BDDs with a uniform variable order. For any permutation of these variables, the number of the subfunctions does not change because they are still represented uniquely with a different but uniform variable order. Therefore, $width_k$ never varies for any permutation of $\{x_1, x_2, \ldots, x_k\}$. □

3.3.2 Algorithm

Our method uses the width of BDDs to estimate the complexity of the graph, and the variables are chosen one by one from the highest to the lowest according to the cost function. As shown in Fig. 3.5, we choose x_k that gives the minimum $width_{k-1}$. We

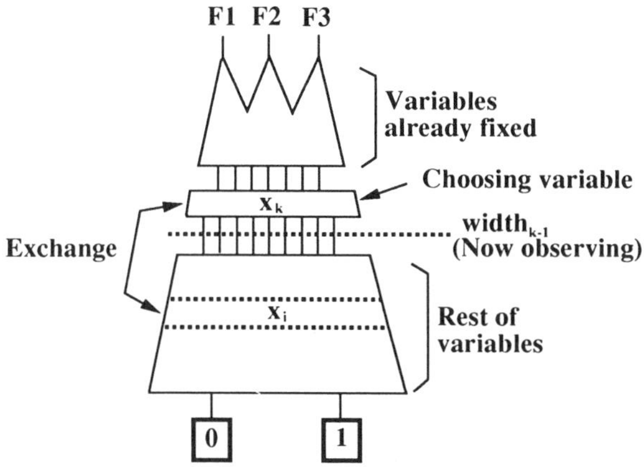

Figure 3.5 Minimum-width method.

call this algorithm *the minimum-width method.* If there are two candidates with the same $width_{k-1}$, we choose the one at the higher position in the initial order.

It is reasonable to use the width as a cost function because:

- On choosing x_k; $width_{k-1}$ is independent of the order of the lower variables $x_1, x_2, \ldots, x_{k-1}$. Therefore, it is not sensitive to variation of the initial order.
- We had better avoid making $width_{k-1}$ large because $width_{k-1}$ is a lower bound of the number of the nodes at the lower than x_k.
- It is not difficult to compute $width_k$.

The reordering of variables is carried out by repeating the exchange of a pair of variables. The exchange between x_i and x_j can be completed by the following sequence of logic operations:

$$f_{ex} = (\overline{x_i} \cdot \overline{x_j} \cdot f_{00}) \vee (\overline{x_i} \cdot x_j \cdot f_{10}) \vee (x_i \cdot \overline{x_j} \cdot f_{01}) \vee (x_i \cdot x_j \cdot f_{11}),$$

where f_{00}, f_{01}, f_{10}, and f_{11} are the subfunctions obtained by assigning a value 0/1 to the two variables x_i and x_j as follows:

$$\begin{array}{lll} f_{00}: & x_i = 0 & x_j = 0 \\ f_{01}: & x_i = 0 & x_j = 1 \\ f_{10}: & x_i = 1 & x_j = 0 \\ f_{11}: & x_i = 1 & x_j = 1. \end{array}$$

This operation requires no traverse of the part of the graph lower than x_i and x_j. The operation time is proportional to the number of the nodes at a higher position than x_i and x_j. Therefore, the higher variables can be exchanged more quickly.

The $width_k$ can be computed by counting the number of subfunctions obtained by assigning any combination of values $\{0, 1\}$ to the variables $x_{k+1}, x_{k+2}, \ldots, x_n$. By traversing the nodes at positions higher than x_k, we can compute the $width_k$ in a time proportional to the number of the visited nodes.

Roughly speaking, the time complexity of our method is $O(n^2G)$, where G is the average size of BDDs during the ordering process. This complexity is considerably less than that of conventional algorithms seeking exactly the best order.

3.3.3 Experimental Results

We implemented the ordering method described above, and conducted some experiments for an evaluation. We used a SPARC Station 2 (SunOS 4.1.1, 32 M Byte), and the program is written in C and C++. The memory requirement for BDDs is about 21 Bytes per node.

In our experiments, we generated initial BDDs for given logic circuits in a certain order of variables and applied our ordering method to the initial BDDs. We used *negative edges.*

The results for some examples are summarized in Table 3.2. In this table, "sel8," "enc8," "add8," and "mult6" are the same ones used in Section 2.5. The other items were chosen from the benchmark circuits in DAC'86[dG86]. In general, these circuits have multiple outputs. Our program handles multiple output functions by using the shared BDD technique. the performance of the reordering method greatly depends on the initial order. We generated 10 initial BDDs in random orders, and applied our ordering method in each case. The table lists the maximum, minimum, and average numbers of the nodes before and after ordering. "Time (sec)" is the average ordering time for the 10 BDDs (including the time needed for generating initial BDDs).

Table 3.2 Results of using the minimum-width method.

Function	In.	Out.	BDD nodes: Ave.(Min.-Max.) Before		After		Time (sec)
sel8	12	2	81.4	(24-217)	22.8	(21-25)	0.30
enc8	9	4	29.8	(27-32)	26.2	(25-28)	0.24
adder8	17	9	557.7	(352-953)	41.0	(41-41)	1.19
mult6	6	6	2963.6	(2486-3363)	2353.7	(2278-2429)	13.07
5xp1	7	10	72.9	(44-87)	41.0	(41-41)	0.44
9sym	9	1	24.0	(24-24)	24.0	(24-24)	0.50
alupla	25	5	8351.3	(4386-13161)	1120.5	(933-1306)	28.47
vg2	25	8	891.8	(539-1489)	91.5	(88-97)	3.35

Table 3.3 Results for large-scale examples.

Function	In.	Out.	BDD nodes Before	After	Time (sec)
c432	36	7	27302	1361	188.3
c499	41	32	52369	40288	1763.6
c880	60	26	23369	9114	874.6
c1908	33	25	17129	8100	252.3
c5315	178	123	31229	2720	6389.3

The results show that our method can reduce the size of BDDs remarkably for most of the examples, except for "9sym," which is a symmetric function. Note that our method constantly gives good results despite the various initial orders.

Similar experiments were then done using larger examples. The functions were chosen from the benchmark circuits in ISCAS'85[BF85]. These circuits are too large and too complicated to generate initial BDDs in a random order, so to obtain a good initial order, we used the DWA method, which is described in Section 3.2.

As shown in Table 3.3, our method is also effective for large-scale functions. It takes longer time, but it is still faster than methods that seek exactly the best order. The sizes of the BDDs after reordering are almost equal to those obtained by using the heuristic methods based on the circuit information[MWBSV88, FFK88, MIY90, BRKM91]. Our results are useful for evaluating other heuristic methods of variable ordering.

Table 3.4 Comparison with incremental search methods.

Function	BDD nodes			
	Initial order	Min-Width	Local search	Combination
sel8 (good)	16	19	16	19
(bad)	510	22	278	22
(random)	81.4	22.8	31.7	20.7
adder8 (good)	41	41	41	41
(bad)	1777	41	1206	41
(random)	557.7	41.0	317.8	41.0
alupla (good)	1037	1082	1037	1047
(bad)	13161	1306	3605	1300
(random)	8351.3	1120.5	3761.0	1087.1

Weak points of our method are that it takes more time than the heuristic methods using the circuit information and that it requires a BDDs with a certain initial order.

We conducted another experiment to compare the properties of the minimum-width method and the incremental search method. We implemented a method that exchanges a pair of variables next to each other if the exchange reduces the size of the BDDs. For three examples, we applied both ordering methods to the same function for various initial orders: the best and worst ones and 10 random orders. The results listed in Table 3.4 show that the two ordering methods have complementary properties. The incremental search never gives results worse than any initial order, but the effect greatly depends on the initial order. The minimum-width method, on the other hand, does not guarantee a result better than the initial order, but the results are constantly close to the best one.

These results led us to conclude that it is effective to apply the minimum-width method at first, and then apply the incremental search for the final optimization. The results of our experiments with such a combination of methods are summarized in Table 3.4, and show that the combination is more effective than either of the methods alone.

3.4 CONCLUSION

We have discussed properties of variable ordering, and have shown two heuristic methods: the DWA method and the minimum-width method. The former one finds an appropriate order before generating BDDs. It refers to the topological information of the Boolean expression or logic circuit that specifies the sequence of logic operations. Experimental results show that the DWA method finds an acceptable order in a short computation time for many practical circuits. The minimum-width method, on the other hand, finds an appropriate order for a given BDD without using additional information. In many cases, this method gives better results than the DWA method in a longer but still reasonable time.

This evaluation of the variable ordering indicates that it is effective to apply these heuristic methods in the following manner:

1. First generate BDDs with an initial order given by a topology-based heuristic, such as the DWA method.
2. To reduce the size of BDDs, apply an exchange-based heuristic with a global sense, such as the minimum-width method.
3. Do the final optimization by using an incremental local search.

This sequence gives fairly good results for many practical problems, but because these methods are only heuristics, there are cases in which they give poor results. Tani et al.[THY93] has proved that an algorithm finding the best order is in the class of NP complete. This implies that it is almost impossible to have a method of variable ordering that always finds best order in a practical time. We will therefore make do with some heuristic methods selected according to the applications.

The techniques of variable ordering are still being studied intensively, and one noteworthy method is the *dynamic variable ordering*, recently presented by Rudell[Rud93]. It is based on having the BDD package itself determine and maintain the variable order. Every time the BDDs grow to a specified size, the reordering process is invoked automatically, just as the garbage collection. This method is very effective in reducing BDD size, although it sometimes takes a long computation time.

4

REPRESENTATION OF MULTI-VALUED FUNCTIONS

In many practical applications related to digital system design, it is a basic technique to use ternary-valued functions containing *don't care*. In this chapter, we discuss the methods for representing *don't care* by using BDDs. These methods can be extended to represent multi-valued logic functions which deal with integer values.

4.1 BOOLEAN FUNCTIONS WITH DON'T CARE

A Boolean function with *don't care* is regarded as a function from a Boolean vector input to a ternary-valued output, denoted as:

$$f : \{0, 1\}^n \rightarrow \{0, 1, d\},$$

where d means *don't care*. Such a function is also called an *incompletely specified Boolean function*. In the following sections, we simply call such a function *ternary-valued function.*

Ternary-valued functions are manipulated by the extended logic operations, and the rules of the logic operations between two ternary-valued functions are defined as follows:

NOT

f	$\overline{f}$
0	1
1	0
d	d

AND

$f \cdot g$	0	1	d
0	0	0	0
1	0	1	d
d	0	d	d

OR

$f \vee g$	0	1	d
0	0	1	d
1	1	1	1
d	d	1	d

EXOR

$f \oplus g$	0	1	d
0	0	1	d
1	1	0	d
d	d	d	d

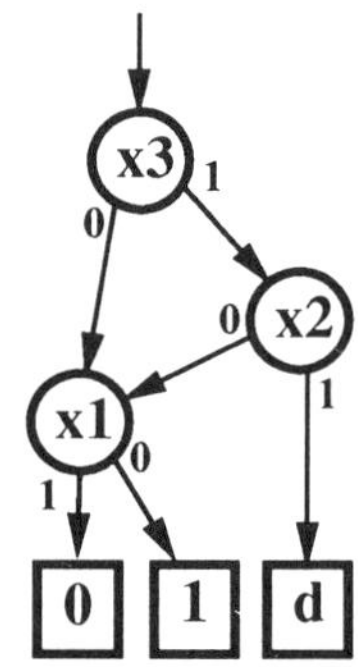

Figure 4.1 A Ternary-valued BDD.

We also define two special unary operations $\lceil f \rceil$ and $\lfloor f \rfloor$ that are important since they are used for abstracting Boolean functions from ternary-valued functions:

f	$\lceil f \rceil$	$\lfloor f \rfloor$
0	0	0
1	1	1
d	1	0

In the operations of the ternary-valued functions, we sometimes refer to a constant function that always returns d. We call it *chaos function.*

4.1.1 Ternary-Valued BDDs and BDD Pairs

There are two ways to represent ternary-valued functions by using BDDs. The first one is to introduce ternary values '0', '1' and 'd' at the terminal nodes of BDDs, as shown in Fig. 4.1. We call this BDD *ternary-valued BDD*. This method is natural and easy to understand, but it has a disadvantage that the operations $\lceil f \rceil$ and $\lfloor f \rfloor$ are not easy and that we have to develop a new BDD package for ternary-valued functions. Matsunaga et al. reported work[MF89] that uses such a BDD package.

The second way is to encode the ternary-value into a pair of Boolean values, and represent a ternary-valued function by a pair of Boolean functions, denote as:

$$f : [f_0, f_1].$$

Since each component f_0 and f_1 can be represented by an conventional BDD, we do not have to develop another BDD package. This idea was first presented by Bryant[CB89], and here we use the encoding[MIY90]:

$$\begin{array}{ll} 0: & [0,0] \\ 1: & [1,1] \\ d: & [0,1], \end{array}$$

which differs from Bryant's code. The choice of encoding is important for the efficiency of the operations. In this encoding, f_0 and f_1 express the functions $\lfloor f \rfloor$ and $\lceil f \rceil$, respectively. Under this encoding, the constant functions 0 and 1 are respectively expressed as $[0,0]$ and $[1,1]$. The *chaos* function is represented as $[0,1]$, and the logic operations are simply computed as follows:

$$\overline{[f_0, f_1]} = [\overline{f_1}, \overline{f_0}],$$

$$[f_0, f_1] \cdot [g_0, g_1] = [f_0 \cdot g_0, f_1 \cdot g_1],$$
$$[f_0, f_1] \vee [g_0, g_1] = [f_0 \vee g_0, f_1 \vee g_1].$$

To find out which method is more efficient, the ternary-valued BDDs or the BDD pairs, we compare them by introducing the *D-variable.*

4.1.2 D-variable

We propose to use a special variable, which we call the *D-variable*, for representing ternary-valued functions. As shown in Fig. 4.2(a), a pair of BDDs $f : [f_0, f_1]$ can be joined into a single BDD using D-variable on the root node whose 0- and 1-edges are pointing to f_0 and f_1, respectively. This BDD has the following elegant properties:

- The constant functions 0 and 1 are represented by the 0- and 1-terminal nodes, respectively. The *chaos* function is represented by a BDD which consists of only one non-terminal node with the D-variable.
- When the function f returns 0 or 1 for any inputs; that is, when f does not contain *don't care*, f_0 and f_1 become the same function and the subgraphs are completely shared. In such cases, the D-variable is redundant and is removed automatically. Consequently, if f contains no *don't care*, its form becomes the same as that of a usual BDD.

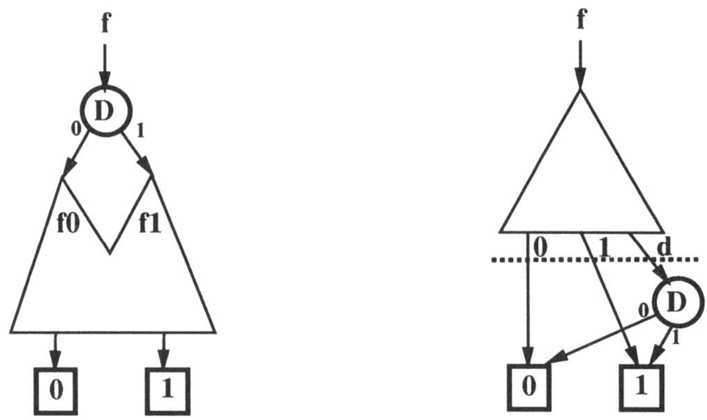

(a) At the highest position. (b) At the lowest position.

Figure 4.2 D-variable.

Using D-variable, we can discuss the relation of the two representation: the ternary-valued BDDs and the BDD pairs. In Fig 4.2(a), the D-variable is ordered at the highest position in the BDD. When the D-variable is reordered to the lowest position, the form of the BDD changes as shown in Fig. 4.2(b). In this BDD, each path from the root node to the 1-terminal node through the D-variable node represents an input assignment that makes $f_0 = 0$ and $f_1 = 1$, namely $f = d$ *(don't care)*. The other paths not through the D-variable node represent the assignments such that $f = 0$ or $f = 1$. (Notice that there are no assignments to have $f_0 = 1$ and $f_1 = 0$.) Therefore, if we regard the D-variable node as a terminal, this BDD corresponds to the ternary-valued BDD.

Consequently, we can say that both the ternary-valued BDDs and the BDD pairs are the special forms of the BDDs using the D-variable, and we can compare the efficiency of the two methods by considering of the properties of variable ordering. From the discussion in previous chapter, we can conclude that the D-variable should be ordered at the higher position when the D-variable greatly affects the function.

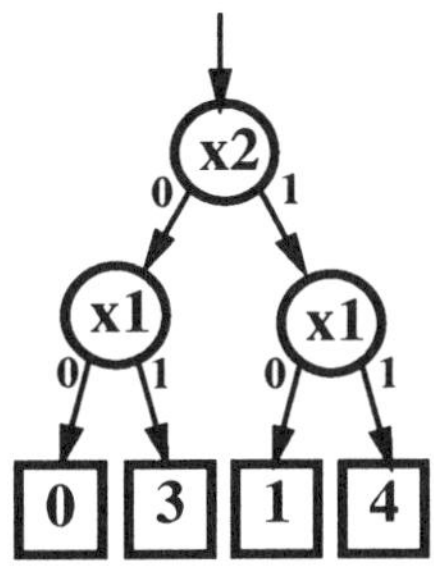

Figure 4.3 A multi-terminal BDD (MTBDD).

4.2 REPRESENTATION OF BOOLEAN-TO-INTEGER FUNCTIONS

Extending the argument about ternary-valued functions, we can represent multi-valued logic functions by using BDDs. In this section, we deal with the functions from Boolean-vector input to an integer output, denoted

$$f : \{0, 1\}^n \rightarrow I.$$

Here we call such functions *Boolean-to-Integer (B-to-I) functions*. Similarly to the ternary-valued functions, there are two ways to represent B-to-I functions using BDDs: *Multi-Terminal BDDs (MTBDDs)* and *BDD vectors[Min93a]*.

MTBDDs are extended BDDs with multiple terminal nodes, each of which has an integer value (Fig. 4.3). This method is natural and easy to understand, but we need to develop a new BDD package to manipulate multi-terminals. Hachtel and Somenzi et al. have reported several investigations of MTBDD[BFG$^+$93, HMPS94]. They call MTBDDs in other words, *Algebraic Decision Diagrams (ADDs)*.

Using BDD vectors is a way to represent B-to-I functions by a number of usual BDDs. By encoding the integer numbers into n-bit binary codes, we can decompose a B-to-I function into n pieces of Boolean functions each of which represents a digit of the binary-corded number. These Boolean functions can then be represented with shared BDDs (Fig. 4.4). This method was mentioned in the work[CMZ$^+$93].

Here we discuss which representation is more efficient in terms of size. We show two extreme examples as follows:

B-to-I function

x1	x2	f (f2 f1 f0)
0	0	0 (0 0 0)
0	1	1 (0 0 1)
1	0	3 (0 1 1)
1	1	4 (1 0 0)

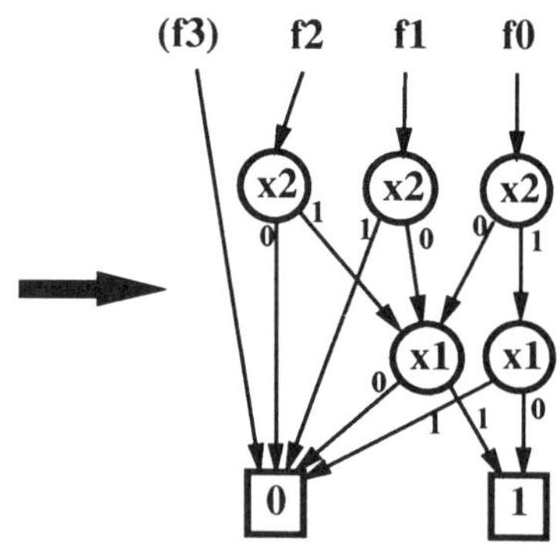

Figure 4.4 A BDD vector.

1. Assume an MTBDD with a large number of non-terminal nodes and a number of terminal nodes with random values of n-bit integers (Fig 4.5(a)).
   If we represent the same function by using an n-bit BDD vectors, these BDDs can hardly be shared with each other since they represent random values (Fig 4.5(b)). In this case, the BDD vector requires about n times as many nodes as the MTBDD does.

2. Assume a B-to-I function for $(x_1 + 2x_2 + 4x_3 + \ldots + 2^{n-1}x_n)$.
   This function can be represented with an n-node BDD vector (Fig 4.6(a)). when using an MTBDD, on the other hand, we need 2^n terminal nodes (Fig 4.6(b)).

Similar to the way we did for the ternary-valued functions, we show that the comparison between multi-terminal BDDs and BDD vectors can be reduced to the variable-ordering problem. Assume the BDD shown in Fig. 4.7(a), which was obtained by combining the BDD vector shown in Fig. 4.4 with what we call *bit-selection variables.* If we change the variable order to move the bit-selection variables from higher to lower positions, the BDD becomes one like that shown in Fig. 4.7(b). In this BDD, the subgraphs with bit-selection variables correspond to the terminal nodes in the MTBDD. That is, MTBDDs and BDD vectors can be transformed into each other by changing the variable order by introducing the bit-selection variables. The efficiency of the two representations thus depends on the nature of the objective functions, and we therefore cannot determine which representation is generally more efficient.

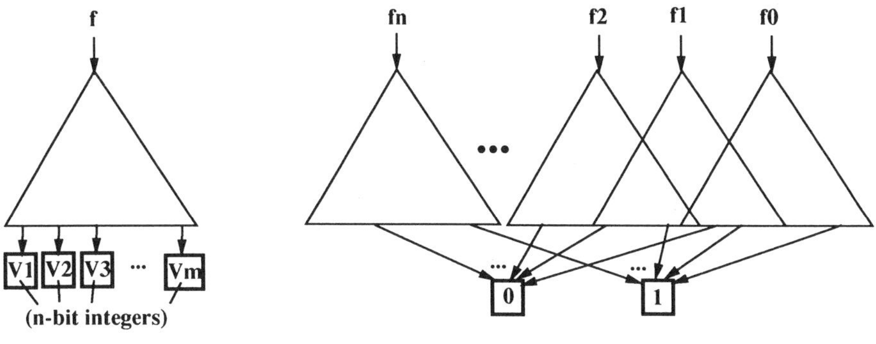

(a) MTBDD. (b) BDD vector.

Figure 4.5 An example where MTBDD is better.

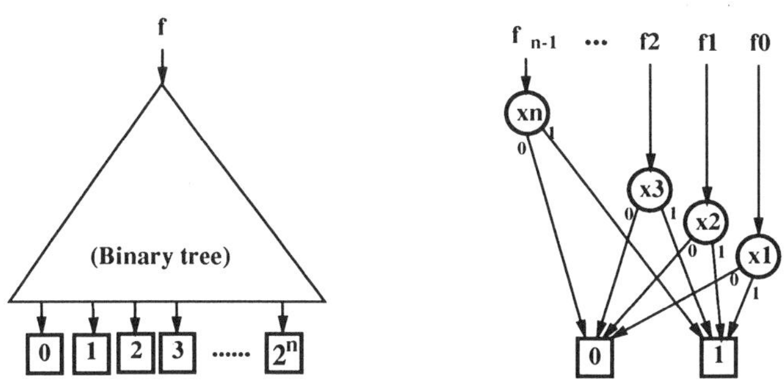

(a) MTBDD. (b) BDD vector.

Figure 4.6 An example where BDD vector is better.

4.3 REMARKS AND DISCUSSIONS

In this chapter, we have discussed the methods for representing multi-valued logic functions. We have shown two methods for handling *don't care* — ternary-valued BDDs and BDD pairs — and we have compared the two by introducing the D-variable. The techniques used for handling *don't care* are basic and important for Boolean function manipulation in many problems. We utilized such techniques effectively for the logic synthesis methods, which is discussed in the following chapters.

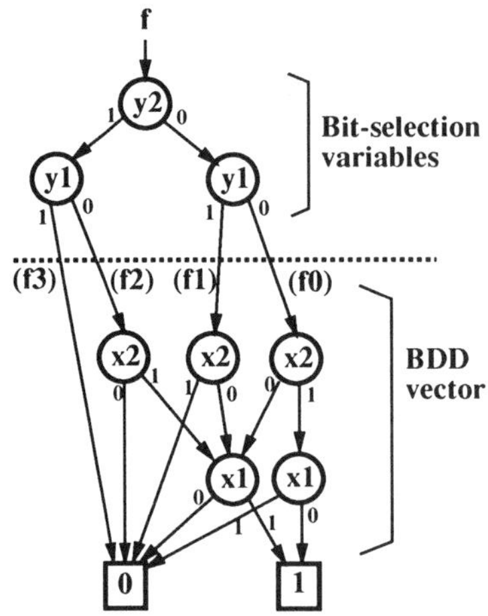

(a) BDD vector with bit-selection variables.

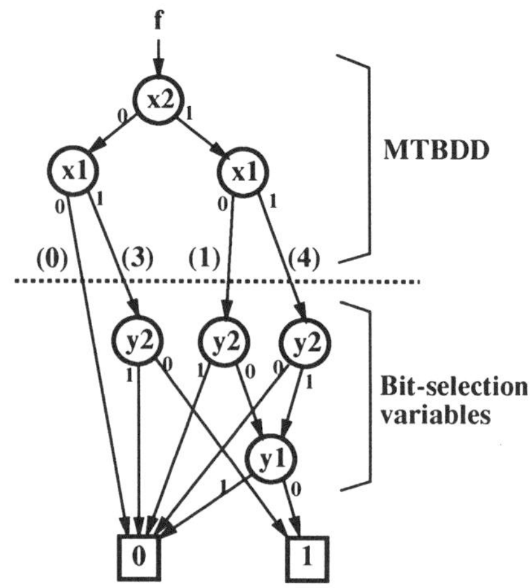

(b) MTBDD with bit-selection variables.

Figure 4.7 Bit-selection variables.

In addition, we extended the arguments to B-to-I functions, and presented two methods: MTBDDs and BDD vectors. These methods can be compared by introducing the bit-selection variables, similarly to the D-variable for the ternary-valued functions. Based on the techniques for manipulating B-to-I functions, we developed an *arithmetic Boolean expression manipulator*, which is presented in Chapter 9.

Several variants of BDDs have recently been devised in order to represent multi-valued logic functions. The two most notable works are the *Edge-Valued BDDs (EVBDDs)* presented by Lai et al.[LPV93] and the *Binary Moment Diagrams (BMDs)* developed by Bryant[BC95]. EVBDDs can be regarded as MTBDDs with attributed edges. The attributed edges indicate that we should add a number to the output values of the functions. Figure 4.8 shows an example of an EVBDD representing a B-to-I function $(x_1 + 2x_2 + 4x_3)$. This technique is sometimes effective to reduce the memory requirement, especially when representing B-to-I functions for linear expressions.

BMDs provide the representation of arithmetic expressions by using a structure similar to that of MTBDDs. In a usual MTBDD, each path from the root node to a terminal node corresponds to an assignment to the input variables, and each terminal node has an output value for the assignment. In a BMD, on the other hand, each path corresponds to a product term in the algebraic expression and each terminal node has

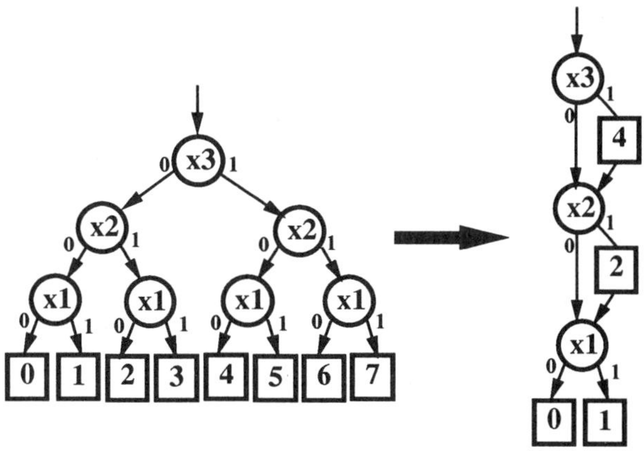

Figure 4.8 EVBDD for $(x_1 + 2x_2 + 4x_3)$.

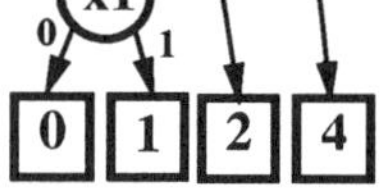

Figure 4.9 BMD for $(x_1 + 2x_2 + 4x_3)$.

a coefficient of the product term. For example, a B-to-I function for $(x_1 + 2x_2 + 4x_3)$ becomes a binary tree form of MTBDD, but the algebraic expression contains only three terms, and it can be represented by a BMD as shown in Fig. 4.9.

Multi-valued logic manipulation is important to broaden the scope of BDD applications. Presently, a number of studies on this topic are in progress. These techniques are useful not only for VLSI CAD but also for various areas in computer science.

5

GENERATION OF CUBE SETS FROM BDDS

In many problems on digital system design, cube sets (also called covers, PLA forms, sum-of-products forms, or two-level logics) are used to represent Boolean functions. They have been extensively studied for many years, and their manipulation algorithms are important in LSI CAD systems. In general, it is not so difficult to generate BDDs from cube sets, but there are no efficient methods for generating compact cube sets from BDDs.

In this chapter, we present a fast method for generating prime-irredundant cube sets from BDDs[Min92a, Min93b]. Prime-irredundant means that each cube is a prime implicant and no cube can be eliminated.

The minimization or optimization of cube sets has received much attention, and a number of efficient algorithms, such as *MINI*[HCO74] and *ESPRESSO* [BHMSV84], have been developed. Since these methods are based on cube set manipulation, they cannot be applied to BDD operations directly. Our method is based on the idea of the *Recursive operator*, proposed by Morreale[Mor70]. We found that Morreale's algorithm can be improved and efficiently adapted for BDD operations.

The features of our method are summarized as follows:

- A prime and irredundant representation can be obtained quickly.
- It generates cube sets from BDDs *directly* without temporarily generating redundant cube sets in the process.
- It can handle *don't cares*.
- The algorithm can be extended to manage multiple output functions.

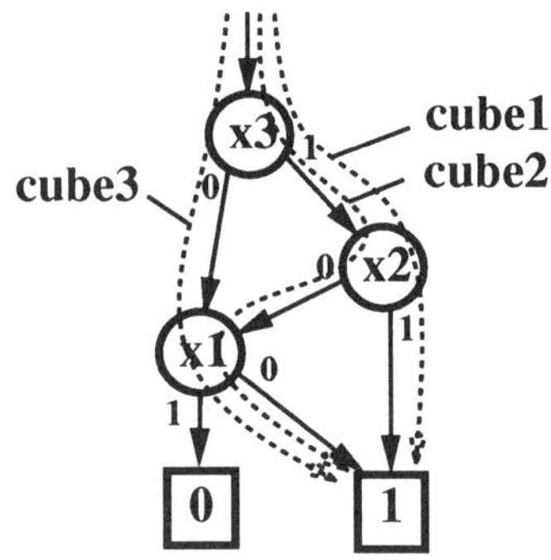

Figure 5.1 1-path enumeration method.

In the remainder of this chapter, we first survey a conventional method for generating cube sets from BDDs, and next we present our algorithm to generate prime-irredundant cube sets. We then show experimental results of our method, followed by conclusion.

5.1 PREVIOUS WORKS

Akers[Ake78] presented a simple method for generating cube sets from BDDs by enumerating the 1-paths. This method enumerates all the paths from the root node to the 1-terminal node, and lists the cubes which correspond to the assignments of the input variables to activate such paths. In the example shown in Fig. 5.1, we can find the three paths that lead to the cube set:

$$(x_3 \cdot x_2) \vee (x_3 \cdot \overline{x_2} \cdot \overline{x_1}) \vee (\overline{x_3} \cdot \overline{x_1}).$$

In reduced BDDs, all the redundant nodes are eliminated, so the literals of the eliminated nodes never appear in the cubes. In the above example, the first cube contains neither x_1 nor $\overline{x_1}$. All of the cubes generated in this method are disjoint because no two paths can be activated simultaneously. But although this method can generate disjoint cube sets, it does not necessarily give the *minimum* ones. For example, the literal of the root node appears in every cube, but some of them may be unnecessary. Considerable redundancy, in terms of the number of cubes or literals, remains in general.

Jacobi and Trullemans[JT92] recently presented a method for removing such redundancy. It generates a prime-irredundant cube set from a BDD in a divide-and-conquer manner. On each node of the BDD, it generates two cube sets for the two subgraphs of the node, and then it combines the two by eliminating redundant literals and cubes. In this method, a cube set is represented with a list of BDDs each of which represents a cube. The redundancy of each cube is determined by applying BDD operations. Although this method can generate compact cube sets, it temporarily generates the lists of redundant cubes during the procedure, and the manipulation of such lists sometimes requires a lot of computation time and space.

5.2 GENERATION OF PRIME-IRREDUNDANT CUBE SETS

In this section, we discuss the properties of prime-irredundant cube sets and present the algorithm for generating such cube sets *directly* from given BDDs.

5.2.1 Prime-Irredundant Cube Sets

If a cube set has the following two properties, we call it a *prime-irredundant cube set.*

- Each cube is a *prime implicant*; that is, no literal can be eliminated without changing the function.
- There are no redundant cubes; that is, no cube can be eliminated without changing the function.Irredundant

The expression $xyz \vee x\overline{y}$, for example, is not prime-irredundant because we can eliminate a literal without changing the function, whereas the expression $xz + \overline{x}\,\overline{y}$ is a prime-irredundant cube set.

Prime-irredundant cube sets are very compact in general, but they are not always the minimum form. The following three expressions represent the same function and all of them are prime-irredundant:

$$x\overline{y} \vee xz \vee \overline{x}y \vee \overline{xz}$$
$$x\overline{y} \vee \overline{x}y \vee yz \vee \overline{yz}$$
$$x\overline{y} \vee \overline{xz} \vee yz$$

We can thus see that prime-irredundant cube sets do not provide unique forms and that the number of cubes and literals can be different. Empirically, however, they are not much larger than the minimum form.

Prime-irredundant cube sets are useful for many applications including logic synthesis, fault testable design, and combinatorial problems.

5.2.2 Morreale's Algorithm

Our method is based on the *recursive operator* proposed by Morreale[Mor70]. His algorithm recursively deletes redundant cubes and literals from a given cube set. The basic idea is summarized in this expansion:

$$isop = (\overline{v} \cdot isop_0) \vee (v \cdot isop_1) \vee isop_d,$$

where $isop$ represents a prime-irredundant cube set and v is one of the input variables. This expansion means that the cube set can be divided into three subsets containing $\overline{v}$, v, and neither. Then when $\overline{v}$ and v are eliminated from each cube, the three subsets $isop_1, isop_0$, and $isop_d$ should also be prime-irredundant. On the basis of this expansion, the algorithm generates a prime-irredundant cube set recursively (see [Mor70] for details).

Unfortunately, Morreale's method is not efficient for large-scale functions because the algorithm is based on cube set representation and it takes a long time to manipulate cube sets for tautology checking, inverting, and other logic operations. However, the basic idea of "recursive expansion" is well suited to BDD manipulation, which is what motivated us to improve and adapt Morreale's method for BDD representation.

5.2.3 ISOP Algorithm Based on BDDs

Our method generates a compact cube set directly from a given BDD, not through redundant cube sets. The algorithm, called *ISOP (Irredundant Sum-Of-Products generation)*, is described in Fig. 5.2, and here we illustrate how it works by using the example shown in Figures 5.3. In Fig. 5.3(a), the function f is divided into the two subfunctions, f_0 and f_1, by assigning $0, 1$ to the input variable ordered at the highest position in the BDD. In Fig. 5.3(b), f_0' and f_1' are derived from f_0 and f_1 by assigning a *don't care* value to the minterms that commonly give $f_0 = 1$ *and* $f_1 = 1$. f_0' and f_1' represent the minterms to be covered by the cubes including v or $\overline{v}$. We thereby generate their prime-irredundant cube sets $isop_0$ and $isop_1$, recursively. In Fig. 5.3(c), f_0'' and f_1'' are derived from f_0 and f_1 by assigning a *don't care* value to the minterms

$ISOP(f(\mathbf{x}))$ {
/* (input) $f(\mathbf{x}) : \{0,1\}^n \rightarrow \{0,1,d\}$ */
/* (output) $isop$: prime-irredundant cube sets*/
  if ($\forall \mathbf{x} \in \{0,1\}^n; f(\mathbf{x}) \neq 1$) { $isop \leftarrow \mathbf{0}$; }
  else if ($\forall \mathbf{x} \in \{0,1\}^n; f(\mathbf{x}) \neq 0$) { $isop \leftarrow \mathbf{1}$; }
  else {
    $v \leftarrow$ one of $\mathbf{x}$;
      /* v is the input with highest order in BDD */
    $f_0 \leftarrow f(\mathbf{x}\,|_{v=0})$; /* the subfunction on $v = 0$ */
    $f_1 \leftarrow f(\mathbf{x}\,|_{v=1})$; /* the subfunction on $v = 1$ */
    Compute f_0', f_1' in the following rules;

f_0':

$f_1 \backslash f_0$	0	1	d
0	0	1	d
1	0	d	d
d	0	d	d

f_1':

$f_1 \backslash f_0$	0	1	d
0	0	0	0
1	1	d	d
d	d	d	d

    $isop_0 \leftarrow ISOP(f_0')$;
      /* recursively generates cubes including $\overline{v}$ */
    $isop_1 \leftarrow ISOP(f_1')$;
      /* recursively generates cubes including v */
    Let g_0, g_1 be the covers of $isop_0, isop_1$, respectively;
    Compute f_0'', f_1'' in the following rules;

f_0'':

$g_0 \backslash f_0$	0	1	d
0	0	1	d
1	–	d	d

f_1'':

$g_1 \backslash f_1$	0	1	d
0	0	1	d
1	–	d	d

    Compute f_d in the following rule;

f_d:

$f_1'' \backslash f_0''$	0	1	d
0	0	0	0
1	0	1	1
d	0	1	d

    $isop_d \leftarrow ISOP(f_d)$;
      /* recursively generates cubes excluding $\overline{v}, v$ */
    $isop \leftarrow (\overline{v} \cdot isop_0) \vee (v \cdot isop_1) \vee isop_d$;
  }
  return $isop$;
}

Figure 5.2 Algorithm for generating prime-irredundant cube sets.

already covered by $isop_0$ or $isop_1$, and in Fig. 5.3(d) f_d is computed from f_0'' and f_1''. f_d represents the minterms to be covered by the cubes excluding v and $\overline{v}$. We thereby generate its prime-irredundant cube set $isop_d$. Finally, the result of $isop$ can be obtained as the union set of $\overline{v} \cdot isop_0$, $v \cdot isop_1$ and $isop_d$.

Note that although here we use Karnaugh maps for purpose of illustration, in practice the functions are represented and manipulated using BDDs.

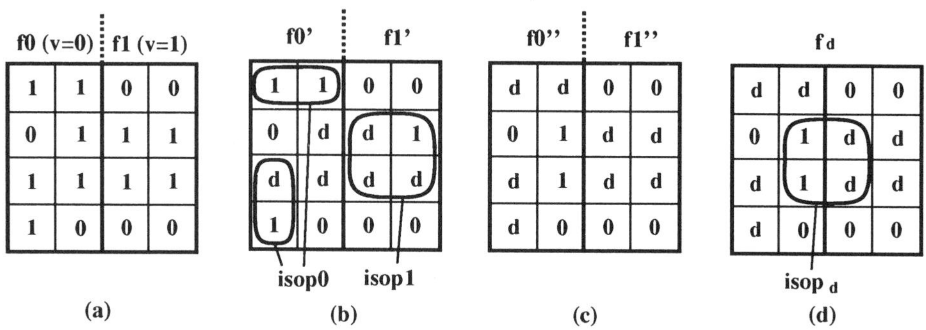

Figure 5.3 An example of using the ISOP algorithm.

If the order of the input variables is fixed, the ISOP algorithm generates a unique form for each function. In other words, it gives a unique form of cube set for a given BDD. Another feature of this algorithm is that it can be applied for functions with don't cares.

This algorithm is well suited to BDD operations because:

- The subfunctions f_0 and f_1 can be derived from f in a constant time.
- Many redundant expansions are avoided automatically because the redundant nodes are eliminated in reduced BDDs.
- BDDs enable fast tautology checking, which is performed frequently in the procedure.

Although it is difficult to precisely evaluate the time complexity of this algorithm. In our experiment, as shown later, the execution time was almost proportional to the product of the initial BDD size and the final cube set size.

5.2.4 Techniques for Implementation

In our algorithm, ternary-valued functions including the *don't care* value are manipulated. As described in Chapter 4.1, we represent them with a pair of binary functions $(\lfloor f \rfloor, \lceil f \rceil)$. In this method, the tautology checking under a *don't care* condition can be written as $\lceil f \rceil \equiv 1$. The special operation for ternary-valued functions can be computed in the combination of ordinary logic operation for $\lfloor f \rfloor$ and $\lceil f \rceil$. For example, the ternary-valued operation

$$f_0' : \begin{array}{c|ccc} {}_{f_1}\backslash^{f_0} & 0 & 1 & d \\ \hline 0 & 0 & 1 & d \\ 1 & 0 & d & d \\ d & 0 & d & d \end{array}$$

can be written as:

$$(\lfloor f_0' \rfloor, \lceil f_0' \rceil) \leftarrow (\lfloor f_0 \rfloor \cdot \overline{\lceil f_1 \rceil},\ \lceil f_0 \rceil).$$

We noted earlier that $isop$ is obtained as the union set of the three parts, as shown in Fig. 5.2. To avoid cube set manipulation, we implemented the method in such a way that the results of cubes are directly dumped out to a file. On each recursive call, we push the processing literal to a stack, which we call a *cube stack*. When a tautology function is detected, the current content of the cube stack is appended to the output file as a cube. This approach is efficient because we manipulate only BDDs, no matter how large the result of the cube set becomes.

Our method can be extended to manage multiple output functions. By sharing the common cubes among different outputs, we obtain a representation more compact than we would if each output were processed separately. In our implementation, the cube sets of all the outputs are generated concurrently; that is, we extend f to be an array of BDDs in order to represent a multiple output function. Repeating recursive calls in the same manner as for a single output function eventuates in the detection of a multiple output constant consisting of 0's and 1's. The 1's mean that corresponding output functions include the cube that is currently kept in the *cube stack*.

5.3 EXPERIMENTAL RESULTS

We implemented the method described in the foregoing section, and conducted some experiments to evaluate its performance. We used a SPARC Station 2 (SunOS 4.1.1, 32 MByte). The program is written in C and C++.

5.3.1 Comparison with ESPRESSO

We first generated initial BDDs for the output functions of practical combinational circuits which may be multi-level or multiple output circuits. We then generated

Table 5.1 Comparison with ESPRESSO.

Func.	In.	Out.	Our method			ESPRESSO		
			Cubes	Literals	Time(s)	Cubes	Literals	Time(s)
sel8	12	2	17	90	0.3	17	90	0.2
enc8	9	4	17	56	0.2	15	51	0.3
add4	9	5	135	819	0.7	135	819	1.9
add8	17	9	2519	24211	13.3	2519	24211	443.1
mult4	8	8	145	945	1.4	130	874	5.0
mult6	12	12	2284	22274	26.7	1893	19340	1126.2
achil8p	24	1	8	32	0.2	8	32	2.0
achil8n	24	1	6561	59049	8.7	6561	59049	3512.7
5xp1	7	10	72	366	0.8	65	347	1.5
9sym	9	1	148	1036	0.9	87	609	10.7
alupla	25	5	2155	26734	20.5	2144	26632	257.3
bw	5	28	68	374	1.1	22	429	1.4
duke2	22	29	126	1296	3.2	87	1036	28.8
rd53	5	3	35	192	0.3	31	175	0.5
rd73	7	3	147	1024	1.2	127	903	4.2
sao2	10	4	76	575	1.1	58	495	2.4
vg2	25	8	110	914	1.9	110	914	42.8
c432	36	7	84235	969037	1744.8	×	×	>36k
c880	60	26	114299	1986014	1096.6	×	×	>36k

prime-irredundant cube sets from the BDDs and counted the numbers of the cubes and literals. We used the DWA method, described in Section 3.2, to find the proper order of the input variables for the initial BDD. The computation time includes the time to determine ordering, the time to generate the initial BDDs, and the time to generate prime-irredundant cube sets. We compared our results with a conventional cube-based method. We flattened the given circuits into cube sets with the system MIS-II[BSVW87], and then optimized the cube sets by using ESPRESSO[BHMSV84].

The results are listed in Table 5.1. The circuits were an 8-bit data selector "sel8," an 8-bit priority encoder "enc8," a (4+4)-bit adder "add4," an (8+8)-bit adder "add8," a (2×2)-bit multiplier "mult4," a (3×3)-bit multiplier "mult6," a 24 input *Achilles' heel function*[BHMSV84] "achil8p," and its complement "achil8n." Other items were chosen from benchmarks at MCNC'90.

Table 5.2 Effect of variable ordering.

Func.	Heuristic order				Random order			
	BDD nodes	Cubes	Literals	Time (sec)	BDD nodes	Cubes	Literals	Time (sec)
sel8	16	17	90	0.3	41	17	90	0.3
enc8	21	17	56	0.2	25	17	56	0.2
add8	41	2519	24211	13.3	383	2519	24211	24.3
mult6	1274	2284	22274	26.7	1897	2354	22963	30.2
achil8n	24	6561	59049	8.7	771	6561	59049	30.9
5xp1	43	72	366	0.8	60	72	364	0.9
alupla	1376	2155	26734	20.4	4309	2155	26730	43.1
bw	85	68	374	1.1	90	64	353	1.1
duke2	396	126	1296	3.2	609	125	1280	3.7
sao2	143	76	575	1.1	133	76	571	1.0
vg2	108	110	914	1.9	1037	110	914	2.7

The table shows that our method is much faster than ESPRESSO — for large-scale circuits, more than 10 times faster. The speed-up was most impressive for the "c432" and "c880," where we generated a prime-irredundant cube set consisting of more than $100,000$ cubes and $1,000,000$ literals within a reasonable time. We could not apply ESPRESSO to these circuits because we were unable to flatten them into cube sets even after ten hours. In another example, ESPRESSO performed poorly for "achil8n" because the *Achilles' heel function* requires a great many cubes when we invert it. Our method nonetheless performed well for "achil8n" because the complementary function can be represented by the same size of BDD as the original one. In general, our method may give somewhat more cubes and literals than ESPRESSO does. In most cases, the differences ranged between 0% and 20%.

5.3.2 Effect of Variable Ordering

We conducted another experiment to evaluate the effect of variable ordering. In general, the size of BDDs greatly depends on the order. We generated prime-irredundant cube sets from the two BDDs of the same function but in a different order: one was in fairly good order (obtained using the *minimum-width method* shown in Section 3.3), and the other was in a random order.

Table 5.3 Result for variation of the number of inputs.

(100 random function, single output)

In.	BDD nodes	Cubes	Literals	Lit./Cubes
1	0.58	0.77	1.35	1.75
2	1.41	1.25	2.84	2.27
3	3.22	2.30	7.17	3.12
4	6.39	4.20	16.05	3.82
5	11.71	7.85	36.39	4.64
6	20.51	14.88	82.18	5.52
7	36.24	27.09	172.06	6.35
8	64.59	52.27	377.41	7.22
9	118.17	99.31	808.09	8.14
10	210.12	192.26	1738.89	9.04
11	365.04	370.90	3693.49	9.96
12	633.97	722.11	7865.91	10.89
13	1144.12	1406.31	16635.79	11.83
14	2154.49	2752.53	35154.84	12.77
15	4151.45	5393.25	73980.57	13.72

As shown in Table 5.2, the numbers of cubes and literals are almost the same for both orders, but the size of BDDs varied greatly. The results demonstrate that our method is robust for variation in order, although variable ordering is still important because it affects the execution time and memory requirement.

5.3.3 Statistical Properties

Taking advantage of our method, we examined the statistical properties of prime-irredundant cube sets. We applied our method to 100 patterns of random functions and calculated the average sizes of the initial BDDs and of the generated cube sets. The random functions were computed using a standard C library.

As shown by the results listed in Table 5.3, the numbers of both BDDs and cubes grow exponentially when the number of inputs is increased. It is known that the maximum BDD size is theoretically $O(2^n/n)$ (where n is the input number)[Ake78], and our statistical experiment produced similar results. In terms of number of cubes, we observe about $O(2^n)$, and the ratio of cubes to literals (the number of literals per cube) is almost proportional to n.

Table 5.4 Result for variation of the number of outputs.

(Number of inputs = 10)

Out.	BDD nodes	Cubes	Literals	Lit./Cubes
1	209.80	192.13	1737.84	9.05
2	364.44	381.69	3452.20	9.04
3	500.86	568.10	5145.01	9.06
4	630.93	754.88	6842.25	9.06
5	758.33	933.86	8468.70	9.07
6	884.87	1120.83	10166.36	9.07
7	1011.08	1294.84	11750.90	9.08
8	1136.94	1471.63	13355.59	9.08
9	1262.29	1649.47	14978.33	9.08
10	1388.76	1815.44	16493.02	9.08
11	1513.15	1987.56	18078.64	9.10

Table 5.4 shows the results obtained when varying the number of outputs while the number of inputs is fixed. Both BDDs and cube sets grow a little less proportionally, thus reflecting sharing the effect of their subgraphs or cubes. We expect such data sharing to be more effective for practical circuits, where the multiple output functions are closely related to each other. The ratio of cubes to literals is almost constant, since the number of inputs is fixed.

We also investigated the relation between this method's performance and the truth table density, which is the rate of 1's in the truth table. We applied our method to the weighted random functions with 10 inputs ranging from 0% to 100% in density. Figure 5.4 shows that the BDD size is symmetric with a center line at 50%, which is like the entropy of information. The number of cubes is not symmetric and peeks at about 60%; however, the number of literals becomes symmetric. This result suggests that the number of literals is better as a measure of the complexity of Boolean functions than is the number of cubes.

5.4 CONCLUSION

We have described the ISOP algorithm for generating prime-irredundant cube sets directly from given BDDs. The experimental results show that our method is much faster than conventional methods. It enables us to generate compact cube sets from

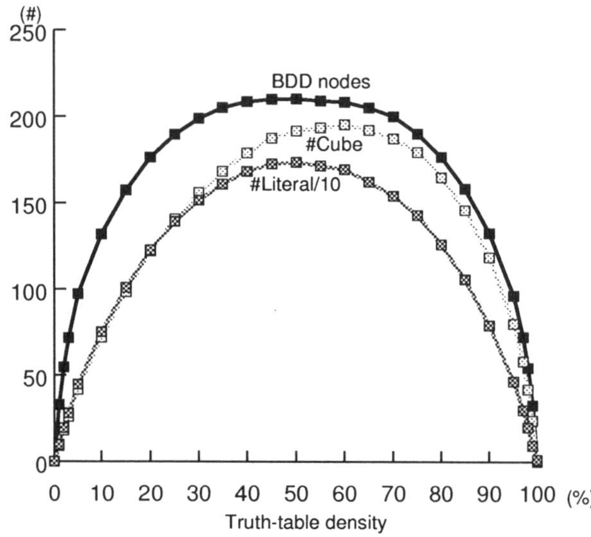

(inputs = 10, output = 1)

Figure 5.4 Result for variation of truth-table density.

large-scale circuits, some of which have not been flattened into cube sets by using the conventional methods. In terms of size of the result, the ISOP algorithm may give somewhat larger results than ESPRESSO, but there are many applications in which such an increase is tolerable. Our method can be used to transform BDDs into compact cube sets or to flatten multi-level circuits into two-level circuits.

Cube set representation sometimes requires an extremely large expression which cannot be reduced any more, while the corresponding BDD is quite small. (An n-bit parity function is a good example.) In such cases, our method can generate cube sets as well, but it is hard to use such large-scale cube sets for practical applications. In the following chapters, a compressed representation of cube sets is presented. It allows us to deal with large-scale cube sets efficiently, and the ISOP algorithm can be accelerated still more.

6 ZERO-SUPPRESSED BDDS

Recently, BDDs have attracted much attention because they enable us to manipulate Boolean functions efficiently in terms of time and space. There are many cases that the algorithm based on conventional data structures can be significantly improved by using BDDs[MF89, BCMD90].

As our understanding of BDDs has deepened, their range of applications has broadened. In VLSI CAD problems, we are often faced with manipulating not only Boolean functions but also *sets of combinations*. By mapping a set of combinations into the Boolean space, they can be represented as a characteristic function by using a BDD. This method enables us to implicitly manipulate a huge number of combinations, which has never been practical before. Two-level logic minimization methods based on implicit set representation have been developed recently[CMF93], and those techniques for manipulating sets of combinations are also used to solve general covering problems[LS90]. Although BDD-based set representation is generally more efficient than the conventional methods, it can be inefficient at times because BDDs were originally designed to represent Boolean functions.

In this chapter, we propose a new type of BDD that has been adapted for set representation[Min93d]. This idea, called a *Zero-suppressed BDD (ZBDD)*, enables us to represent sets of combinations more efficiently than using conventional BDDs. We also discuss *unate cube set algebra*[Min94], which is convenient for describing ZBDD algorithms or procedures. We present efficient methods for computing unate cube set operations, and show some practical applications of these methods.

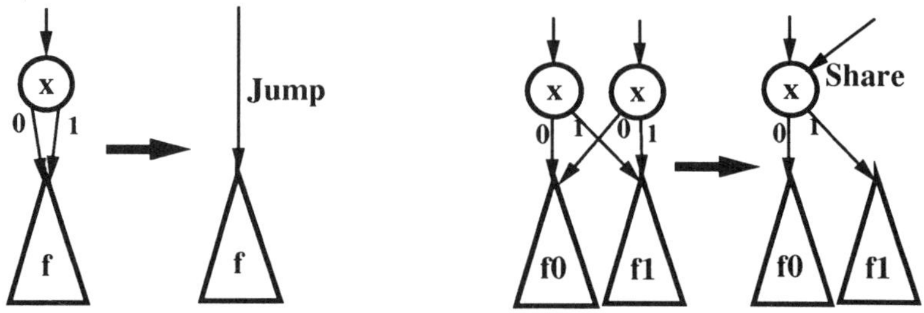

(a) Node elimination. (b) Node sharing.

Figure 6.1 Reduction rules of conventional BDDs.

6.1 BDDS FOR SETS OF COMBINATIONS

Here we examine the reduction rules of BDDs when applying them to represent sets of combinations. We then show a problem which motivates us to develop a new type of BDDs.

6.1.1 Reduction Rules of BDDs

As mentioned in Chapter 2, BDDs are based on the following reduction rules:

1. Eliminate all the redundant nodes whose two edges point to the same node. (Fig. 6.1(a))
2. Share all the equivalent subgraphs. (Fig. 6.1(b))

BDDs give canonical forms for Boolean functions when the variable ordering is fixed, and most uses of BDDs are based on the above reduction rules.

It is important how BDDs are shrunk by the reduction rules. One recent paper[LL92] shows that, for general (or random) Boolean functions, node sharing makes a much more significant contribution to storage saving than the node elimination. For practical functions, however, the node elimination is also important. For example, as shown in Fig. 6.2, the form of a BDD does not depend on the number of input variables as long as the expressions of the functions are the same. When we use BDDs, the irrelevant variables are suppressed automatically and we do not have to consider them. This is

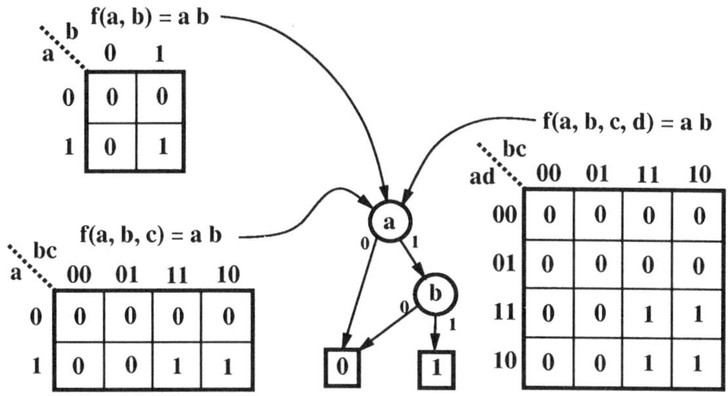

Figure 6.2 Suppression of irrelevant variables in BDDs.

significant because sometimes we manipulate a function that depends on only a few variables among hundreds. This suppression of the irredundant variables is due to the node elimination of BDDs.

6.1.2 Sets of Combinations

Presently, there have been many works on BDD applications, but some of them do not use BDDs to simply represent Boolean functions. we are often faced with manipulating *sets of combinations*. Sets of combinations are used for describing solutions to combinatorial problems. We can solve combinatorial problems by manipulating sets of combinations. The representation and manipulation of sets of combination are important techniques for many applications.

A combination of n items can be represented by an n-bit binary vector, $(x_n x_{n-1} \dots x_2 x_1)$, where each bit, $x_k \in \{1, 0\}$, expresses whether or not the corresponding item is included in the combination. A set of combinations can be represented by a set of the n-bit binary vectors. Sets of combinations can be regarded as subsets of the power set on n items.

We can represent a set of combinations with a Boolean function by using n-input variables for each bit of the vector. The output value of the function expresses whether or not each combination specified by the input variables are included in the

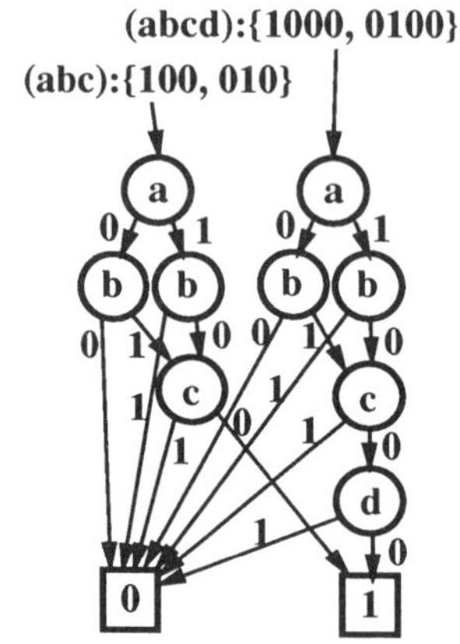

Figure 6.3 BDDs representing sets of combinations.

set. Such Boolean functions are called *characteristic functions*. The set operations such as union, intersection, and difference can be executed by logic operations on characteristic functions.

By using BDDs for characteristic functions, we can manipulate sets of combination efficiently. In such BDDs, the paths from the root node to the 1-terminal node, which we call *1-paths*, represent possible combinations in the set. Because of the effect of node sharing, BDDs compactly represent sets of combinations with a huge number of elements. In many practical cases, the size of graphs becomes much less than the number of elements. BDDs can be generated and manipulated within a time that is roughly proportional to the size of graphs, whereas the operations of the previous set representations (such as arrays and sequential lists) require a time proportional to the number of elements.

Despite the efficiency of manipulating sets by using BDDs, there is one inconvenience: that, as shown in Fig. 6.3, the form of BDDs depends on the number of input variables. We therefore have to fix the number of input variables before generating BDDs. This inconvenience comes from the difference in the model on default variables. In sets of combinations, irrelevant objects never appear in any combination, so default variables are regarded as zero when the characteristic function is true. Unfortunately, such variables can not be suppressed in the BDD representation. We therefore have to generate many useless nodes for irrelevant variables when we manipulate very sparse

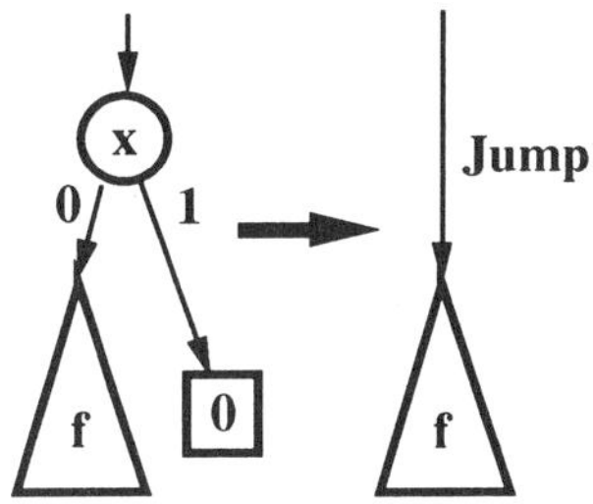

Figure 6.4 New reduction rule for ZBDDs.

combinations. In such cases, the node elimination does not work well in reducing the graphs.

In the following section, we describe a method that solves this problem by using BDDs based on new reduction rules.

6.2 ZERO-SUPPRESSED BDDS

To represent sets of combinations efficiently, we propose the following reduction rules:

1. Eliminate all the nodes whose 1-edge points to the 0-terminal node and use the subgraph of the 0-edge, as shown in Fig. 6.4.
2. Share all equivalent subgraphs in the same way as for ordinary BDDs.

Notice that, contrary to the rules for ordinary BDDs, we do not eliminate the nodes whose two edges point to the same node. This reduction rule is asymmetric for the two edges, as we do not eliminate the nodes whose 0-edge points to a terminal node.

We call BDDs based on the above rules *Zero-suppressed BDDs (ZBDDs).* If the number and the order of input variables are fixed, a ZBDD uniquely represents a Boolean function. This is obvious because a non-reduced binary tree can be reconstructed from a ZBDD by applying the reduction rule reversely.

Figure 6.5 shows ZBDDs representing the same sets of combinations shown in Fig. 6.3. A feature of ZBDDs is that the form is independent of the number of inputs as long

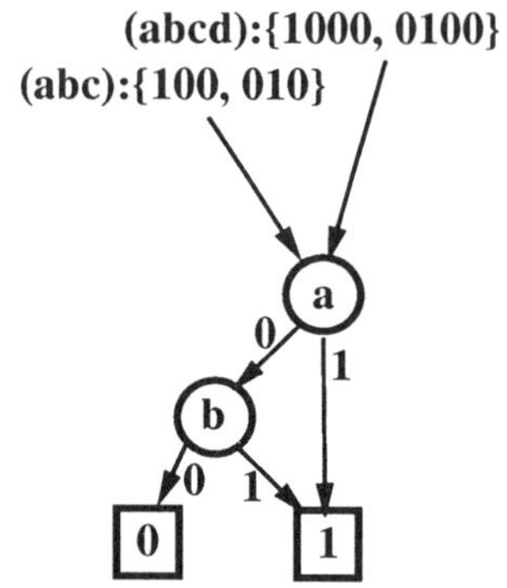

Figure 6.5 ZBDDs representing sets of combinations.

as the sets of combinations are the same. Thus we do not have to fix the number of input variables before generating graphs. ZBDDs automatically suppress the variables which never appear in any combination. It is very efficient when we manipulate very sparse combinations.

To evaluate the efficiency of ZBDDs, we conducted a statistical experiment. We generated a set of one hundred combinations each of which selects k out of 100 objects randomly. We then compared the sizes of the ZBDDs and conventional BDDs representing these random combinations. The result in varying k (Fig. 6.6) shows that ZBDDs are much more compact than conventional ones — especially when k is small. This means that ZBDDs are particularly effective for representing sets of sparse combinations. The effect weakens for large k; however, we can use complement combinations to make k small. For example, the combination selecting 90 out of 100 objects is equivalent to selecting the remaining 10 objects.

Another advantage of ZBDDs is that the number of 1-paths in the graph is exactly equal to the number of combinations in the set. In conventional BDDs, the node elimination breaks this property, so we conclude that ZBDDs are more suitable for representing sets of combinations than conventional BDDs are.

On the other hand, it would be better to use conventional BDDs when representing ordinary Boolean functions, as shown in Fig. 6.2. The difference is in the models of default variables: "fixed to zero" in sets of combinations, and "both the same" in Boolean functions. We can choose one of the two types of BDDs according to the feature of applications.

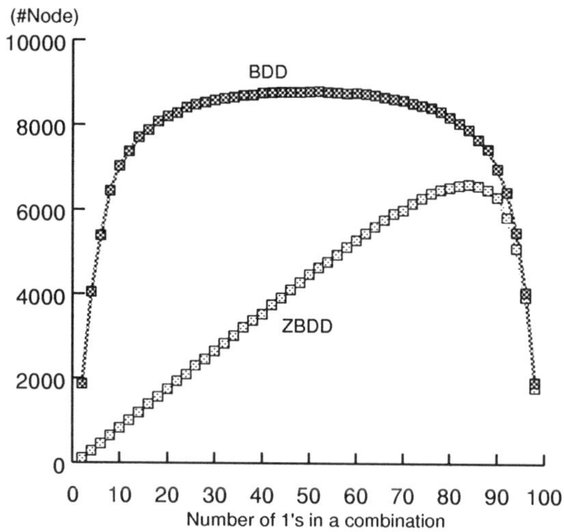

Figure 6.6 Comparison of BDDs and ZBDDs.

6.3 MANIPULATION OF ZBDDS

In this section, we show that ZBDDs are manipulated efficiently as well as conventional BDDs.

6.3.1 Basic Operations

In generating conventional BDDs, we first generate BDDs with only one input variable for each input, and then we construct more complicated BDDs by applying logic operations, such as AND, OR, and EXOR. ZBDDs are also constructed from the trivial graphs by applying basic operations, but the kinds of operations are different since ZBDDs are adapted for sets of combinations.

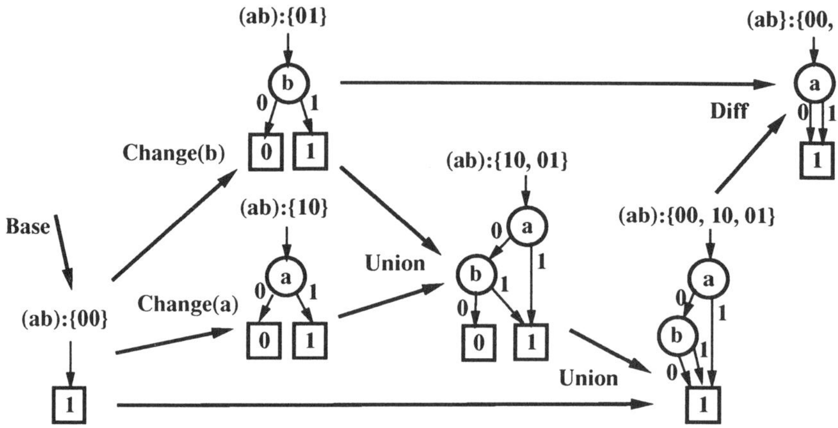

Figure 6.7 Generation of ZBDDs.

We arranged the following line up of basic operations for ZBDDs:

Empty()	returns ϕ. (empty set)
Base()	returns $\{\epsilon\}$.
Subset1(P, var)	returns the subset of P such as $var = 1$.
Subset0(P, var)	returns the subset of P such as $var = 0$.
Change(P, var)	returns P when var is inverted.
Union(P, Q)	returns $(P \cup Q)$
Intsec(P, Q)	returns $(P \cap Q)$
Diff(P, Q)	returns $(P - Q)$
Count(P)	returns $\|P\|$. (number of combinations)

Figure 6.7 shows examples of ZBDDs generated with those operations. Empty() returns the 0-terminal node, and Base() is the 1-terminal node. Any combination can be generated with Base() operation followed by Change() operations for all the variables that appear in the combination. Using Intsec() operation, we can check whether a combination is contained in a set.

In ZBDDs, we do not have the NOT operation, which is an essential operation in conventional BDDs. Its absence here is reasonable, since the complement set $\overline{P}$ cannot be computed if the universal set U is not defined. Using the difference operation, $\overline{P}$ can be computed as $(U - P)$.

6.3.2 Algorithms

We show here that the basic operations for ZBDDs can be executed recursively, like the ones for conventional BDDs.

First, to describe the algorithms simply, we define the procedure Getnode(top, P_0, P_1), which generates (or copies) a node for a variable top and two subgraphs P_0, P_1. In the procedure, we use a hash table, called *uniq-table*, to keep each node unique. Node elimination and sharing are managed only by Getnode( ).

```
Getnode (top, P0, P1) {
    if (P1 == φ) return P0;  /* node elimination */
    P = search a node with (top, P0, P1) in uniq-table;
    if (P exist) return P;   /* node sharing */
    P = generate a node with (top, P0, P1);
    append P to the uniq-table;
    return P;
}
```

Using Getnode( ), we can describe the operations for ZBDDs as follows. Here $P.top$ means a variable with the highest order, and P_0, P_1 are the two subgraphs.

```
Subset1 (P, var) {
   if (P.top < var) return φ;
   if (P.top == var) return P1;
   if (P.top > var)
      return Getnode(P.top, Subset1(P0, var), Subset1(P1, var));
}

Subset0(P, var) {
   if (P.top < var) return P;
   if (P.top == var) return P0;
   if (P.top > var)
      return Getnode(P.top, Subset0(P0, var), Subset0(P1, var));
}

Change (P, var) {
   if (P.top < var) return Getnode(var, φ, P);
   if (P.top == var) return Getnode(var, P1, P0);
   if (P.top > var)
      return Getnode(P.top, Change(P0, var), Change(P1, var));
}
```

```
Union (P, Q) {
  if (P == φ) return Q;
  if (Q == φ) return P;
  if (P == Q) return P;
  if (P.top > Q.top) return Getnode(P.top, Union(P0, Q), P1);
  if (P.top < Q.top) return Getnode(Q.top, Union(P, Q0), Q1);
  if (P.top == Q.top)
    return Getnode(P.top, Union(P0, Q0), Union(P1, Q1));
}

Intsec (P, Q) {
  if (P == φ) return φ;
  if (Q == φ) return φ;
  if (P == Q) return P;
  if (P.top > Q.top) return Intsec(P0, Q);
  if (P.top < Q.top) return Intsec(P, Q0);
  if (P.top == Q.top)
    return Getnode(P.top, Intsec(P0, Q0), Intsec(P1, Q1));
}

Diff (P, Q) {
  if (P == φ) return φ;
  if (Q == φ) return P;
  if (P == Q) return φ;
  if (P.top > Q.top) return Getnode(P.top, Diff(P0, Q), P1);
  if (P.top < Q.top) return Diff(P, Q0);
  if (P.top == Q.top)
    return Getnode(P.top, Diff(P0, Q0), Diff(P1, Q1));
}

Count (P) {
  if (P == φ) return 0;
  if (P == {0}) return 1;
  return Count(P0) + Count(P1);
}
```

In the worst case, these algorithms take an exponential time for the number of variables, but we can accelerate them by using a cache which memorizes the results of recent operations in the same manner as it is used in conventional BDDs. By referring to the cache before every recursive call, we can avoid duplicate executions for equivalent subgraphs. This enable us to execute these operations in a time that is roughly proportional to the size of the graphs.

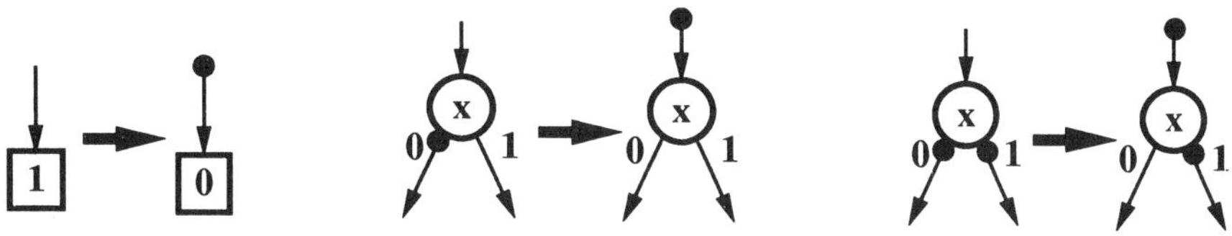

Figure 6.8 Rules for 0-element edges.

6.3.3 Attributed Edges

In conventional BDDs, we can reduce the execution time and memory requirement by using *attributed edges*[MIY90] to indicate certain logic operations such as inverting. ZBDDs also have a kind of attributed edges, but the functions of the attributes are different from conventional ones.

Here we present an attributed edge for ZBDDs. This attributed edge, called *0-element edge*, indicates that the pointing subgraph has a 1-path that consists of 0-edges only. In other word, a 0-element edge means that the set includes the null-combination "ϵ." We use the notation $\hat{P}$ to express the 0-element edge pointing to P.

As with other attributed edges, we have to place a couple of constraints on the location of 0-element edges in order to keep the uniqueness of the graphs:

- Use the 0-terminal only, since $\{\epsilon\}$ can be written as $\hat{\phi}$
- Do not use 0-element edges at the 0-edge on each node.

If necessary, 0-element edges can be carried over as shown in Fig. 6.8. The constraint rules can be implemented in Getnode( ).

0-element edges accelerate operations on ZBDDs. For example, the result of Union(P, $\{\epsilon\}$) depends only on whether or not P includes the "ϵ." In such a case, we can get the result in a constant time when using 0-element edges; otherwise we have to repeat the expansion until P becomes a terminal node.

6.4 UNATE CUBE SET ALGEBRA

In this section, we discuss unate cube set algebra for manipulating sets of combinations. A cube set consists of a number of cubes, each of which is a combination of literals. *Unate* cube sets allow us to use only positive literals, not the negative ones. Each cube represents one combination, and each literal represents an item chosen in the combination .

We sometimes use cube sets to represent Boolean functions, but they are usually *binate* cube sets containing both positive and negative literals. Binate cube sets have different semantics from unate cube sets. In binate cube sets, literal x and $\overline{x}$ represent $x = 1$ and $x = 0$, respectively, and the absence of a literal means *don't care*; that is, $x = 1, 0$, both OK. In unate cube sets, on the other hand, literal x means $x = 1$ and the absence of a literal means $x = 0$. For example, the cube set expression $(a + bc)$ represents $(abc) : \{111, 110, 101, 100, 011\}$ under the semantics of binate cube sets, but $(abc) : \{100, 011\}$ under unate cube set semantics.

6.4.1 Basic Operations

Unate cube set expressions consist of trivial sets and algebraic operators. There are three kinds of trivial sets:

$$\begin{array}{ll} 0 & \text{(empty set),} \\ 1 & \text{(unit set),} \\ x_k & \text{(single literal set).} \end{array}$$

The unit set "1" includes only one cube that contains no literals. This set becomes the unit element of the product operation. A single literal set x_k includes only one cube that consists of only one literal. In the rest of this section, a lowercase letter denotes a literal, and an uppercase letter denotes an expression.

We arranged the line-up of the basic operators as follows:

$$\begin{array}{ll} \& & \text{(intersection),} \\ + & \text{(union),} \\ - & \text{(difference),} \\ * & \text{(product),} \\ / & \text{(quotient of division),} \\ \% & \text{(remainder of division).} \end{array}$$

(We may use a comma "," instead of "+" and we sometimes omit "$*$.") The operation "$*$" generates all possible concatenations of two cubes in respective cube sets. Examples of calculation are:

$$\begin{aligned}
\{ab, b, c\} \& \{ab, 1\} &= \{ab\} \\
\{ab, b, c\} + \{ab, 1\} &= \{ab, b, c, 1\} \\
\{ab, b, c\} - \{ab, 1\} &= \{b, c\} \\
\{ab, b, c\} * \{ab, 1\} &= (ab * ab) + (ab * 1) + (b * ab) \\
&\quad + (b * 1) + (c * ab) + (c * 1) \\
&= \{ab, abc, b, c\}
\end{aligned}$$

There are the following formulas in the unate cube calculation:

$$\begin{aligned}
&P + P = P \\
&a * a = a, \quad (P * P \neq P \ \text{ in general}) \\
&(P - Q) = (Q - P) \Longleftrightarrow (P = Q) \\
&P * (Q + R) = (P * Q) + (P * R)
\end{aligned}$$

Dividing P by Q acts to seek out the two cube sets P/Q (quotient) and $P\%Q$ (remainder) under the equality $P = Q * (P/Q) + (P\%Q)$. In general this solution is not unique. Here we apply the following rules to fix the solution with reference to the *weak-division method*[BSVW87].

1. When Q includes only one cube, (P/Q) is obtained by extracting a subset of P, which consists of the cubes including all the literals in Q's cube, and then eliminating Q's literals. For example,

$$\{abc, bc, ac\}/\{bc\} = \{a, 1\}.$$

2. When Q consists of multiple cubes, (P/Q) is the intersection of all the quotients dividing P by respective cubes in Q. For example,

$$\begin{aligned}
&\{abd, abe, abg, cd, ce, ch\}/\{ab, c\} \\
= \ &(\{abd, abe, abg, cd, ce, ch\}/\{ab\}) \& (\{abd, abe, abg, cd, ce, ch\}/\{c\}) \\
= \ &\{d, e, g\} \& \{d, e, h\} \\
= \ &\{d, e\}.
\end{aligned}$$

3. $(P\%Q)$ can be obtained by calculating $P - Q * (P/Q)$.

These three trivial sets and six basic operators are used to represent and manipulate sets of combinations. In Section 6.3, we defined three other basic operations — Subset1(

), Subset0( ), and Change( ) — for assigning a value to a literal, but we do not have to use them since the weak-division operation can be used as *generalized cofactor* for ZBDDs. For example, Subset1(P, x_k) can be described as (P/x_k), and Subset0(P, x_k) becomes $(P\%x_k)$. And the Change( ) operation can be described by using some multiplication and division operators. Using unate cube set expressions, we can elegantly express the algorithms or procedures for manipulating sets of combinations.

6.4.2 Algorithms

We show here that the basic operations of unate cube set algebra can be efficiently executed using ZBDD techniques. The three trivial cube sets are represented by simple ZBDDs. The empty set "0" becomes the 0-terminal, and the unit set "1" is the 1-terminal node. A single literal set x_k corresponds to the single-node graph pointing directly to the 0- and 1-terminal node. The intersection, union, and difference operations are the same as the basic ZBDD operations shown in Section 6.3. The other three operations — product, quotient, and remainder — are not included in the basic ones, so we have developed algorithms for computing them.

If we calculate the multiplication and division operations by processing each cube one by one, the computation time will depend on the length of expressions. Such a procedure is impractical when we deal with very large numbers of cubes. We developed new recursive algorithms based on ZBDDs in order to calculate large-scale expressions efficiently.

Our algorithms are based on the divide-and-conquer method. Suppose x is the highest-ordered literal. P and Q are then factored into the two parts:

$$P = x * P_1 + P_0, \quad Q = x * Q_1 + Q_0.$$

The product $(P * Q)$ can be written as:

$$(P * Q) = x * (P_1 * Q_1 + P_1 * Q_0 + P_0 * Q_1) + P_0 * Q_0.$$

Each sub-product term can be computed recursively. The expressions are eventually broken down into trivial ones and the results are obtained. In the worst case, this algorithm would require an exponential number of recursive calls for the number of literals, but we can accelerate it by using a hash-based cache that memorizes results of recent operations. By referring to the cache before every recursive call, we can avoid duplicate executions for equivalent subsets. Consequently, the execution time depends on the size of ZBDDs rather than on the number of cubes and literals. This algorithm is shown in detail in Fig. 6.9.

```
procedure(P * Q)
{   if (P.top < Q.top) return (Q * P) ;
    if (Q = 0) return 0 ;
    if (Q = 1) return P ;
    R ← cache("P * Q") ;  if (R exists) return R ;
    x ← P.top ; /* the highest variable in P */
    (P0, P1) ← factors of P by x ;
    (Q0, Q1) ← factors of Q by x ;
    R ← x (P1 * Q1 + P1 * Q0 + P0 * Q1) + P0 * Q0 ;
    cache("P * Q") ← R ;
    return R ;
}
```

Figure 6.9 Algorithm for product.

```
procedure(P/Q)
{   if (Q = 1) return P ;
    if (P = 0 or P = 1) return 0 ;
    if (P = Q) return 1 ;
    R ← cache("P/Q") ;  if (R exists) return R ;
    x ← Q.top ; /* the highest variable in Q */
    (P0, P1) ← factors of P by x ;
    (Q0, Q1) ← factors of Q by x ; /* (Q1 ≠ 0) */
    R ← P1/Q1 ;
    if (R ≠ 0) if (Q0 ≠ 0) R ← R & P0/Q0 ;
    cache("P/Q") ← R ;
    return R ;
}
```

Figure 6.10 Algorithm for division.

Division is computed in the same recursive manner. Suppose that x is a literal at the root-node in Q, and that $P_0, P_1, Q_0, andQ_1$ are the subsets of cubes factored by x. (Notice that $Q_1 \neq 0$, since x appears in Q.) The quotient (P/Q) can be described as

$$(P/Q) = (P_1/Q_1), \quad \text{when} \quad Q_0 = 0,$$
$$(P/Q) = (P_1/Q_1) \mathbin{\&} (P_0/Q_0), \quad \text{otherwise.}$$

Each sub-quotient term can be computed recursively. Whenever we find that one of the sub-quotients $(P_1/Q_1) or (P_0/Q_0)$ results in 0, $(P/Q) = 0$ becomes obvious and we no longer need to compute it. Using the cache technique avoids duplicate

```
***** Unate Cube set Calculator (Ver. 1.1) *****
ucc> symbol a(2) b(1) c(2) d(3) e(2)
ucc> F = (a + b) (c + d + e)
ucc> print F
 a c, a d, a e, b c, b d, b e
ucc> print .factor F
 ( a + b ) ( c + d + e )
ucc> print .matrix F
  1.1..
  1..1.
  1...1
  .11..
  .1.1.
  .1..1
ucc> print .cube F
   6
ucc> print .bddnode F
   5 (10)
ucc> G = F * a + c d e
ucc> print G
 a b c, a b d, a b e, a c, a d, a e, c d e
ucc> print .factor G
 a ( b + 1 ) ( c + d + e ) + c d e
ucc> print F & G
 a c, a d, a e
ucc> print F - G
 b c, b d, b e
ucc> print G - F
 a b c, a b d, a b e, c d e
ucc> print  G / (a b)
 c, d, e
ucc> print  G % (a b)
 a c, a d, a e, c d e
ucc> print .mincost G
 a c (4)
ucc> exit
```

Figure 6.11 Execution of unate cube set calculator.

executions for equivalent subsets. This algorithm is illustrated in Fig. 6.10. The remainder $(P\%Q)$ can be determined by calculating $P - P * (P/Q)$.

6.5 IMPLEMENTATION AND APPLICATIONS

Based on the above techniques, we developed a *Unate Cube set Calculator (UCC)*, which is an interpreter with a lexical and syntax parser for calculating unate cube set expressions by using ZBDDs. Our program allows up to 65,535 different literals. An example of execution is shown in Fig. 6.11.

Using UCC, we can compute the minimum-cost cube under a definition of the cost for each literal. After ZBDDs are constructed, the minimum-cost cube can be found in a time proportional to the number of nodes in the graph, as when using conventional BDDs[LS90].

Because the unate cube set calculator can generate huge ZBDDs with millions of nodes, limited only by memory capacity, we can manipulate large-scale and complicated expressions. Here we show several applications for the unate cube set calculator.

6.5.1 8-Queens Problem

The 8-queens problem is an example in which using unate cube set calculation is more efficient than using ordinary Boolean expressions.

First, we allocate 64 logic variables to represent the squares on a chessboard. Each variable denotes whether or not there is a queen on that square. The problem can be described with the variables as follows:

- In a particular column, only one variable is "1."
- In a particular row, only one variable is "1."
- On a particular diagonal line, one or no variable is "1."

By unate cube set calculation, we can solve the 8-queens problem efficiently. The algorithm can be written as

$$\begin{aligned} S_1 &= x_{11} + x_{12} + \ldots + x_{18} \\ S_2 &= x_{21}(S_1 \% x_{11} \% x_{12}) + x_{22}(S_1 \% x_{11} \% x_{12} \% x_{13}) \\ &\quad + \ldots + x_{28}(S_1 \% x_{17} \% x_{18}) \\ S_3 &= x_{31}(S_2 \% x_{11} \% x_{13} \% x_{21} \% x_{22}) \\ &\quad + x_{32}(S_2 \% x_{12} \% x_{14} \% x_{21} \% x_{22} \% x_{23}) \\ &\quad + \ldots + x_{38}(S_2 \% x_{16} \% x_{18} \% x_{27} \% x_{28}) \\ S_4 &= \ldots \end{aligned}$$

These expressions mean that the strategy is:

Table 6.1 Results on N-queens problems.

N	Lit.	Sol.	BDD nodes	ZBDD nodes	(B/Z)	(Z/S)
4	16	2	29	8	3.6	4.0
5	25	10	166	40	4.2	4.0
6	36	4	129	24	5.4	6.0
7	49	40	1098	186	5.9	4.65
8	64	92	2450	373	6.6	4.05
9	81	352	9556	1309	7.3	3.72
10	100	724	25944	3120	8.3	4.31
11	121	2680	94821	10503	9.0	3.92
12	144	14200	435169	45833	9.5	3.23
13	169	73712	2044393	204781	10.0	2.78

(B/Z): BDD nodes / ZBDD nodes
(Z/S): ZBDD nodes / Solution.

S_1: Search all the choices to put the first queen.

S_2: Search all the choices to put the second queen, considering the first queen's location.

S_3: Search all the choices to put the third queen, considering the first and second queen's location.

...

S_8: Search all the choices to put the eighth queen, considering the other queens' locations.

Calculating these expressions with ZBDDs provides the set of solutions to the 8-queens problem. Okuno[Oku94] reported experimental results for N-queens problems to compare ZBDDs and conventional BDDs. In Table 6.1, the column "BDD nodes" shows the size of BDDs using Boolean algebra, and "ZBDD nodes" shows the size of ZBDDs using unate cube set algebra. We can see that there are about N times fewer ZBDDs than conventional BDDs. We can represent all the solutions at once within a storage space roughly proportional to the number of solutions.

6.5.2 Fault Simulation

Takahashi et al.[TIY91] proposed a fault simulation method considering multiple faults by using BDDs. it is a deductive method for multiple faults, that manipulates

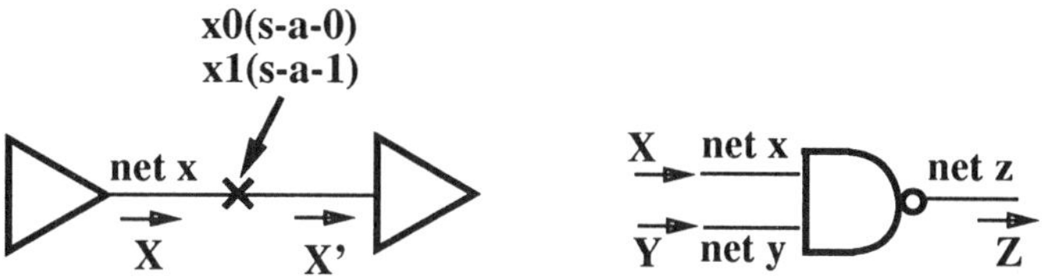

Figure 6.12 Propagation of fault sets

sets of multiple stuck-at faults. It propagates the fault sets from primary inputs to primary outputs, and eventually obtains the detectable faults at primary outputs. Takahashi et al. used conventional BDDs, but we can compute the fault simulation more simply by using ZBDDs based on unate cube set algebra.

First, we generate the whole set of multiple faults that is assumed in the simulation. The set F_1 of all the single stuck-at faults is expressed as

$$F_1 = (a_0 + a_1 + b_0 + b_1 + c_0 + c_1 + ...),$$

where a_0 and a_1 represent the stuck-at-0 and -1 faults, respectively, at the net a. Other literals are expressed similarly. We can represent the set F_2 of double and single faults as $(F_1 * F_1)$. Further more, $(F_2 * F_1)$ gives the set of three or fewer multiple faults. If we assume exactly double (not including single) faults, we can calculate $(F_2 - F_1)$. In this way, the whole set U can easily be described with unate cube set expressions.

After computing the whole set U, we then propagate the detectable fault set from the primary inputs to the primary outputs. As illustrated in Fig. 6.12(a), two faults x_0 and x_1 are assumed at a net x. Let X and X' be the detectable fault sets at the source and sink, respectively, of the net x. We can calculate X' from X with the following unate cube expressions:

$$X' = (X + (U/x_1) * x_1)\%x_0, \text{ when } x = 0 \text{ in a good circuit.}$$
$$X' = (X + (U/x_0) * x_0)\%x_1, \text{ when } x = 1 \text{ in a good circuit.}$$

On each gate, we calculate the fault set at the output of the gate from the fault sets at the inputs of the gate. Let us consider a NAND gate with two inputs x and y, and one output z, as shown in Fig. 6.12(b). Let X, Y, and Z be the fault sets at x, y, and z. We can calculate Z from X and Y by the simple unate cube set operations as follows:

$$Z = X \& Y, \text{ when } x = 0, y = 0, z = 1 \text{ in a good circuit.}$$

$$Z = X - Y, \text{ when } x = 0, y = 1, z = 1 \text{ in a good circuit.}$$
$$Z = X + Y, \text{ when } x = 1, y = 1, z = 0 \text{ in a good circuit.}$$

We can compute the detectable fault sets by calculating those expressions for all the gates in the circuit. Using unate cube set algebra, we can simply describe the fault simulation procedure and can execute it directly by using a unate cube set calculator.

6.6 CONCLUSION

We have proposed ZBDDs, which are BDDs based on a new reduction rule, and have presented their manipulation algorithms and applications. ZBDDs can represent sets of combinations uniquely and more compactly than conventional BDDs. The effect of ZBDDs is remarkable especially when we are manipulating sparse combinations. On the basis of the ZBDD techniques, we have discussed the method for calculating unate cube set algebra, and we have developed a unate cube set calculator, which can be applied to many practical problems.

Unate cube sets have semantics different from those of binate cube sets, but there is a way to simulate binate cube sets by using unate ones: we use two unate literals x_1 and x_0 for one binate literal. For example, a binate cube set $(a\ \overline{b} + c)$ is expressed as the unate cube set $(a_1 b_0 + c_1)$. In this way, we can easily simulate the cube-based algorithms implemented in logic design systems such as ESPRESSO and MIS[BSVW87]. Using this technique, we have developed a practical multi-level logic optimizer (detailed in the next chapter).

Unate cube set expressions are suitable for representing sets of combinations, and they can be efficiently manipulated using ZBDDs. For solving some types of combinatorial problems, our methods are more useful than those using conventional BDDs. We expect the unate cube set calculator to be a helpful tool in developing VLSI CAD systems and in various other applications.

7

MULTI-LEVEL LOGIC SYNTHESIS USING ZBDDS

Logic synthesis and optimization techniques have been used successfully for practical design of VLSI circuits in recent years. Multi-level logic, Optimization is important in logic synthesis systems and a lot of research in this field has been undertaken[MKLC87, MF89, Ish92]. In particular, the *algebraic logic minimization method*, such as MIS[BSVW87], is the most successful and prevalent way to attain this optimization. It is based on cube set (or two-level logic) minimization and generates multi-level logic from cube sets by applying a *weak-division method*. This approach is efficient for functions that can be expressed in a feasible size of cube sets, but we are sometimes faced with functions whose cube set representations grow exponentially with the number of inputs. Parity functions and full-adders are examples of such functions. This is a problem of the cube-based logic synthesis methods.

The use of BDDs provided a break-through for that problem. By mapping a cube set into the Boolean space, a cube set can be represented as a Boolean function using a BDD. Using this method, we can represent a huge number of cubes implicitly in a small storage space. This enables us to manipulate very large cube sets whose manipulation has not been practicable before. Based on the Cube set, BDD-based representation, new cube set minimization methods have been developed [CMF93, MSB93].

As noted in Chapter 6, even though BDD-based cube representation is usually more efficient than the conventional methods, it can sometimes be inefficient because BDDs were originally designed to represent Boolean functions. We have recently developed ZBDDs which are adapted for representing sets of combinations, and they enable us to represent cube sets more efficiently. They are especially effective when we manipulate cube sets using intermediate variables to represent multi-level logic.

In this chapter, we presents a fast weak-division algorithm[Min93c] for implicit cube sets based on ZBDDs. This algorithm can be computed in a time almost proportional to the number of nodes in ZBDDs, which are usually much smaller than the number of literals in the cube set. By using this method, we can quickly generate multi-level logic from cube sets even for parity functions and full-adders, that have not been possible to handle when using the conventional algebraic methods. We implemented a new multi-level logic synthesizer using the implicit weak-division method, and experimental results indicate our method is much faster than conventional methods, especially when larger cube sets are manipulated. The implicit weak-division method is thus expected to significantly accelerate logic synthesis systems and enlarges the scale of the circuits to which these systems are applicable.

The following sections, we first discusses the implicit cube set representation based on ZBDDs. We then present the implicit weak-division method and show experimental results.

7.1 IMPLICIT CUBE SET REPRESENTATION

Cube sets (also called covers, PLAs, sum-of-products forms, and two-level logic) are used to represent Boolean functions in many problems in the design and testing of digital systems. In a cube sets, each cube is formed by a combination of positive and negative literals for input variables. (We are speaking here of a *binate* cube set, different from the *unate* cube set discussed in Chapter 6.) In this section, we present an implicit method for representing cube sets using ZBDDs, and show a method for generating prime-irredundant cube sets using the implicit representation.

7.1.1 Cube Set Representation Using ZBDDs

Coudert and Madre developed a method, called *Meta products*[CMF93], for representing cube sets using BDDs. Meta-products are BDD representations for characteristic functions of cube sets. In their method, two variables are used for each input, and the two variables determine the existence of the literal and whether it is positive or negative. Coudert and Madre also presented further reduced graphs, called *Implicit Prime Sets (IPS)*[CM92], to represent prime cube sets efficiently. However, IPSs can represent only *prime* cube sets and cannot provide canonical expressions for *general* cube sets.

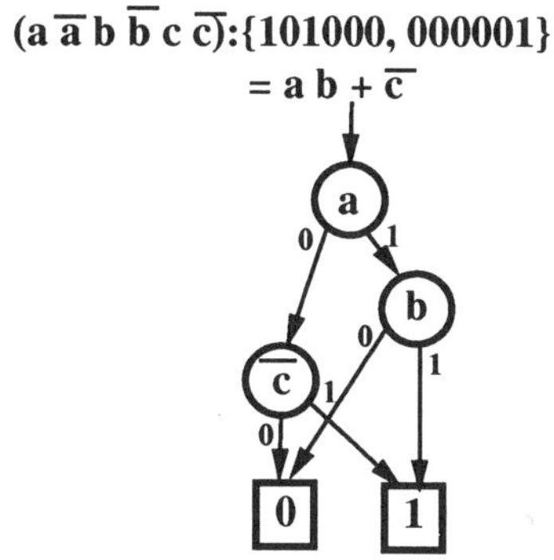

Figure 7.1 Implicit cube set representation based on ZBDDs.

By using ZBDDs, we can represent any cube set simply, efficiently, and uniquely. Figure 7.1 illustrates a cube set that can be seen as a set of combinations using two variables for literals x_k and $\overline{x_k}$. x_k and $\overline{x_k}$ never appear together in the same cube, and at least one should be 0. The 0's are conveniently suppressed in ZBDDs. The number of cubes exactly equals the number of 1-paths in the graph, and the total number of literals can be counted in a time proportional to the size of the graph.

The basic operations for the cube set representation based on ZBDDs are the following:

"0"	returns ϕ. (no cube)
"1"	returns **1**. (the tautology cube)
And0(P, var)	returns $(\overline{var} \cdot P)$.
And1(P, var)	returns $(var \cdot P)$.
Factor0(P, var)	returns the factor of P by $\overline{var}$.
Factor1(P, var)	returns the factor of P by var.
FactorD(P, var)	returns the cubes in P excluding var, $\overline{var}$.
Union(P, Q)	returns $(P + Q)$.
Intsec(P, Q)	returns $(P \cap Q)$.
Diff(P, Q)	returns $(P - Q)$.
CountCubes(P)	returns number of cubes.
CountLits(P)	returns number of literals.

"0" corresponds to the 0-terminal node on ZBDDs, and "1" corresponds to the 1-terminal node. Any cube can be generated by applying a number of And0( ) and And1( ) operations to "1." The three Factor operations mean that

$$P = (\overline{var} \cdot \text{Factor0}) + (var \cdot \text{Factor1}) + \text{FactorD}.$$

Intsec() is different from the logical AND operation; it extracts only the common cubes in the two cube sets. These operations are simply composed of ZBDD operations, and their execution time is roughly proportional to the size of the graphs.

7.1.2 ISOP Algorithm for Implicit Cube Set Representation

Using this new cube set representation, we have developed a program for generating prime-irredundant cube sets based on the *ISOP algorithm*, described in Chapter 5. Our program converts a conventional BDD representing a given Boolean function into a ZBDD representing a prime-irredundant cube set.

The ISOP algorithm is summarized as this expansion:

$$isop = \overline{v} \cdot isop_0 + v \cdot isop_1 + isop_d,$$

where $isop$ represents the prime-irredundant cube set, and v is one of the input variables. This expansion reveals that $isop$ can be divided into three subsets containing $\overline{v}$, v, and neither. When $\overline{v}$ and v are excluded from each cube, the three subsets $isop_1, isop_0$, and $isop_d$ should also be prime-irredundant. Based on this expansion, the algorithm generates a prime-irredundant cube set recursively.

We found that the ISOP algorithm can be accelerated through the use of the new cube set representation based on ZBDDs. We prepared a hash-based cache to store the results of each recursive call. Each entry in the cache is formed as a pair (f, s), where f is the pointer to a given BDD and s is the pointer to the result of the ZBDD. On each recursive call, we check the cache to see whether the same subfunction f has already appeared, and if so, we can avoid duplicate processing and return result s directly. By using this technique, we can execute the ISOP algorithm in a time roughly proportional to the size of graph, independent of number of the cubes and literals.

We implemented the program on a SPARC station 2 (128 MB) and generated prime-irredundant cube sets from the functions of large-scale combinational circuits. The results are listed in Table 7.1. The following circuits were used for this experiment: an (8+8)-bit adder "add8," a (16+16)-bit adder "add16," a (6×6)-bit multiplier "mult6," an (8×8)-bit multiplier "mult8," and a 24-input *Achilles' heel function*[BHMSV84] "achil8p" and its complement "achil8n." Other circuits were chosen from the benchmarks of MCNC'90.

The columns "In." and "Out." show the scale of the circuits. The column "BDD nodes for func." lists the number of nodes in conventional BDDs representing Boolean functions. "Cubes" and "Literals" show the scale of the generated prime-

Table 7.1 Generation of prime-irredundant cube sets.

Circuit size				Result of cube set generation			
Func.	In.	Out.	BDD nodes for func.	Cubes	Literals	ZBDD nodes for cubes	Time (sec)
add8	17	9	41	2,519	23,211	88	0.5
add16	33	17	81	655,287	11,468,595	176	1.3
mult6	12	12	1,358	2,284	22,273	3,315	5.7
mult8	16	16	10,766	35,483	474,488	45,484	82.3
achil8	24	1	24	8	32	24	0.3
achil8n	24	1	24	6,561	59,049	24	0.3
c432	36	7	27,302	84,235	969,028	14,407	83.2
c499	41	32	52,369	348,219,564	6,462,057,445	195,356	2400.1
c880	60	26	19,580	114,299	1,986,014	18,108	78.7
c1908	33	25	17,129	56,323,472	1,647,240,617	233,775	385.6
c5315	178	123	32,488	137,336,131	742,606,419	41,662	886.4

Table 7.2 Effect of ZBDDs for cube set representation.

Func.	ZBDD nodes for cubes	BDD nodes for cubes	Z/B (%)	Density* (%)
add8	88	292	30.1	27.1
add16	176	844	20.8	26.5
mult6	3,315	6,170	53.7	40.6
mult8	45,484	78,802	57.7	41.8
achil8	24	118	20.3	8.3
achil8n	24	78	30.8	18.7
c432	14,407	45,319	31.8	16.0
c499	195,356	342,728	57.0	22.6
c880	18,108	78,444	23.0	14.5
c1908	233,775	404,432	57.8	44.3
c5315	41,662	174,566	23.8	1.5

*Density = (Literals/Cubes) / (In.$\times$ 2)

irredundant cube sets. They indicate the memory requirement when we use a classical representation, such as a linear linked list. The actual memory requirements of the implicit cube set representation are listed in the column "ZBDD nodes for cubes." As we can see, extremely large prime-irredundant cube sets containing billions of literals can easily be generated in a practical computation time. Such cube sets have never been generated before. Our experimental results thus show that the implicit cube set representation reduces the memory requirement dramatically.

We also evaluated the effect of using ZBDDs. In Table 7.2, the column "BDD nodes for cubes" lists the number of nodes needed when we use conventional BDDs to represent implicit cube set representation. "Z/B" is the ratio of size of ZBDDs and conventional BDDs. "Density" shows how many literals appear in a cube on the average. This result indicates the property that ZBDDs are effective for representing sparse combinations. Notice that v and $\overline{v}$ never appear in the same cube, so the density does not exceed 50%. In general, reduced cube sets consist of sparse cubes and the use of ZBDDs is effective. When we manipulate cube sets using intermediate variables to represent multi-level logic, cube sets are sparser, and ZBDDs more effective.

7.2 FACTORIZATION OF IMPLICIT CUBE SET REPRESENTATION

In this section, we present a fast algorithm for factorizing implicit cube representation using ZBDDs, and we demonstrate results of our new multi-level logic optimizer based on the algorithm.

7.2.1 Weak-Division Method

In general, two-level logics can be factorized into more compact multi-level logics. The initial two-level logics are represented with large cube sets for primary output functions, as shown in Fig. 7.2(a). When we determine a good intermediate logic, we make a cube set for it and reduce the other existing cube sets by using a new intermediate variable. Eventually, we construct a multi-level logic that consists of a number of small cube sets, as illustrated in Fig 7.2(b). The multi-level logic consists of hundreds of cube sets, each of which is very small. On the average, less than 10 variables out of hundreds are used for each cube set. They yield so sparse combinations that the use of ZBDDs is quite effective. Another benefit of ZBDDs is that we do not have to fix the number of variables beforehand. We can use additional variables whenever an intermediate logic is found.

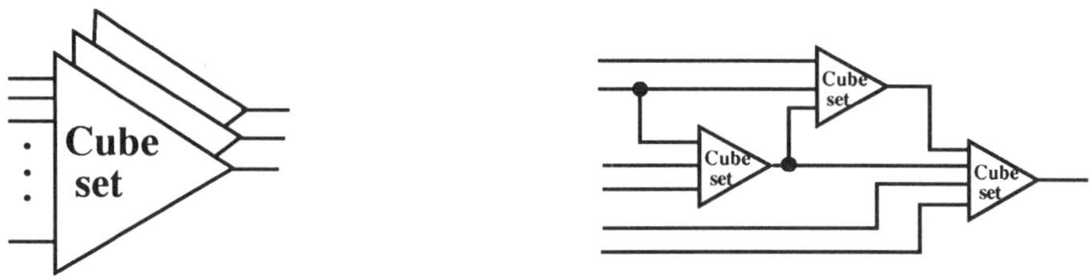

(a) Initial cube sets. (b) Final cube set network.

Figure 7.2 Factorization of cube sets.

Weak-division (or algebraic division) is the most successful and prevalent method for generating multi-level logics from cube sets. For example, the cube set expression

$$f = a\,b\,d + a\,b\,\overline{e} + a\,b\,\overline{g} + c\,d + c\,\overline{e} + c\,h$$

can be divided by $(a\,b + c)$. By using an intermediate variable p, we can rewrite the expression

$$f = p\,d + p\,\overline{e} + a\,b\,\overline{g} + c\,h \quad p = a\,b + c.$$

In the next step, f will be divided by $(d + \overline{e})$ in a similar manner.

Weak-division does not exploit all of Boolean properties of the expression and is only an algebraic method. In terms of result quality, it is not as effective as other stronger optimizing methods, such as the *transduction method*[MKLC87]. However, weak-division is still important because it is used for generating initial logic circuits for other strong optimizers, and is applied to large-scale logics that cannot be handled by strong optimizers.

The conventional weak-division algorithm is executed by computing the common part of quotients for respective cubes in the divisor. For example, suppose the two expressions are

$$\begin{aligned} f &= a\,b\,d + a\,b\,\overline{e} + a\,b\,\overline{g} + c\,d + c\,\overline{e} + c\,h, \\ p &= a\,b + c. \end{aligned}$$

f can be rewritten as:

$$f = a\,b\,(d + \overline{e} + \overline{g}) + c\,(d + \overline{e} + h).$$

The quotient (f/p) can then be computed as

$$(f/p) \quad = \quad (f/\,(a\,b)\,) \cap (f/c)$$

$$
\begin{aligned}
&= (d + \overline{e} + \overline{g}) \cap (d + \overline{e} + h) \\
&= d + \overline{e}.
\end{aligned}
$$

The remainder $(f\%p)$ is computed using the quotient:

$$
\begin{aligned}
(f\%p) &= f - p\,(f/p) \\
&= a\,b\,\overline{g} + c\,h.
\end{aligned}
$$

Using the quotient and the remainder, we can reduce f:

$$
\begin{aligned}
f &= p\,(f/p) + (f\%p) \\
&= p\,d + p\,\overline{e} + a\,b\,\overline{g} + c\,h.
\end{aligned}
$$

One step in the factorization process is then completed.

The conventional weak-division algorithm requires an execution time that depends on the length of expressions (or the number of literals in f and p) because we have to compute a number of quotients for all cubes in the divisor. This method is therefore impracticable when we deal with very large cube sets such as those for parity functions and adders. In the next section, we present a much faster weak-division algorithm based on the implicit cube set representation using ZBDDs.

7.2.2 Fast Weak-Division for Implicit Cube Set Representation

Our method generates (f/p) from f and $p(\neq 0)$ in the implicit cube set representation. The algorithm is described in Fig. 7.3. The basic idea here is that we do not compute quotients for respective cubes in the divisor, but rather for subsets of cubes factored by an input variable. Here, v is the highest-ordered input variable contained in p, and cube sets f and p are factored into three parts as

$$
\begin{aligned}
f &= \overline{v}\,f_0 + v\,f_1 + f_d, \\
p &= \overline{v}\,p_0 + v\,p_1 + p_d.
\end{aligned}
$$

The quotient (f/p) can then be written as

$$
(f/p) = (f_0/p_0) \cap (f_1/p_1) \cap (f_d/p_d).
$$

Each sub-quotient term can be computed recursively. The procedure is eventually broken down into trivial problems and the results are obtained. If one of the values for p_0, p_1, or p_d is zero, we may skip the term. For example, if $p_1 = 0$, then $(f/p) = (f_0/p_0) \cap (f_d/p_d)$. Whenever we find that one of the values for (f_0/p_0),

```
procedure(f/p)
{   if (p = 1) return f ;
    if (f = 0 or f = 1) return 0 ;
    if (f = p) return 1 ;
    q ← cache("f/p") ; if (q exists) return q ;
    v ← p.top ; /* the highest variable in p */
    (f0, f1, fd) ← factors of f by v ;
    (p0, p1, pd) ← factors of p by v ;
    q ← p ;
    if (p0 ≠ 0) q ← f0/p0 ;
    if (q = 0) return 0 ;
    if (p1 ≠ 0)
        if (q = p) q ← f1/p1 ;
        else q ← q ∩ (f1/p1) ;
    if (q = 0) return 0 ;
    if (pd ≠ 0)
        if (q = p) q ← fd/pd ;
        else q ← q ∩ (fd/pd) ;
    cache("f/p") ← q ;
    return q ;
}
```

Figure 7.3 Implicit weak-division algorithm.

(f_1/p_1) and (f_d/p_d) becomes zero, $(f/p) = 0$ becomes obvious and we no longer need to continue the calculation.

This algorithm computes the example shown in the previous section as follows:

$$\begin{aligned}
& (a\,b\,d + a\,b\,\overline{e} + a\,b\,\overline{g} + c\,d + c\,\overline{e} + c\,h)/(a\,b + c) \\
= \; & (b\,d + b\,\overline{e} + b\,\overline{g})/b \;\cap\; (c\,d + c\,\overline{e} + c\,h)/c \\
= \; & (d + \overline{e} + \overline{g})/1 \;\cap\; (d + \overline{e} + h)/1 \\
= \; & (d + \overline{e} + \overline{g}) \cap (d + \overline{e} + h) \\
= \; & d + \overline{e}.
\end{aligned}$$

In the same way as for the ISOP algorithm, we prepared a hash-based cache to store results for each recursive call and avoid duplicate execution. Using the cache technique, we can execute this algorithm in a time nearly proportional to the size of the graph, regardless of the number of cubes and literals.

```
procedure(f · g)
{    if (f.top < g.top) return (g · f) ;
     if (g = 0) return 0 ;
     if (g = 1) return f ;
     h ← cache("f · g") ; if (h exists) return h ;
     v ← f.top ; /* the highest variable in f */
     (f0, f1, fd) ← factors of f by v ;
     (g0, g1, gd) ← factors of g by v ;
     h ← v̄(f0 · g0 + f0 · gd + fd · g0)
         +v(f1 · g1 + f1 · gd + fd · g1) + fd · gd ;
     cache("f · g") ← h ;
     return h ;
}
```

Figure 7.4 Implicit multiplication algorithm.

To obtain the remainder of division $(f\%p) = f - p\ (f/p)$, we need to compute the algebraic multiplication between two cube sets. This procedure can also be described recursively and executed quickly using the cache technique, as illustrated in Fig. 7.4.

7.2.3 Divisor Extraction

For multi-level logic synthesis based on the weak-division method, the quality of the results greatly depends on the choice of divisors. *Kernel extraction*[BSVW87] is the most common and sophisticated method for obtaining divisors. It extracts good divisors and has been used successfully in practical systems such as MIS-II. For very large cube sets, however, this method is complicated and time consuming.

Here we present a simple and fast method for finding divisors in implicit cube sets. The basic algorithm is described as follows:

```
Divisor(f)
{    v ← a literal appears twice in f ;
     if(v exist) return Divisor(f/v) ;
     else return f ;
}
```

If there is a literal that appears more than once in a cube set, we compute the factor for the literal. Repeating this recursively, we eventually obtain a divisor, which is the same as the one called a *Level-0 kernel* in the kernel extraction method used in

MIS-II. With this method, factors for a literal can be computed quickly in the implicit representation. Whether a literal appears more than once can be checked efficiently by looking on the branch of the graph.

A different divisor may be obtained for another order of factoring literals. When two or more possible literals are located, we choose one defined later so that the extracted divisor may have variables nearer to the primary inputs. This rule allows us to maintain a shallow depth of the circuit.

The use of a common divisor for multiple cube sets may yield better results, but locating common divisors is complicated and time consuming for large cube sets. So far, we have been able to extract only single output divisors and apply them to all the other cube sets. When there is a cube set providing a non-zero quotient for the divisor, we factorize the cube set. At least one cube set and sometimes more can be divided by a common cube.

Using the complement function for the divisor, we sometimes can attain more compact expressions. For example,

$$f = a\,\overline{c} + b\,\overline{c} + \overline{a}\,\overline{b}\,c$$

can be factorized using a complement divisor as:

$$\begin{aligned} f &= p\,\overline{c} + \overline{p}\,c, \\ p &= a + b. \end{aligned}$$

It is not easy to compute the complement function in the cube set representation. We transform the cube set into a conventional BDD for the Boolean function of the divisor, and make a complement for the BDD. We then regenerate a cube set from the inverted BDD by using the ISOP algorithm. This strategy seems as if it would require a large computation time, however, the actual execution time is comparatively small in the entire process because the divisors are always small cube sets.

7.3 IMPLEMENTATION AND EXPERIMENTAL RESULTS

Based on the above method, we implemented a multi-level logic synthesizer. The basic flow of the program is illustrated in Fig. 7.5. Starting with non-optimized multi-level logics, we first generate BDDs for the Boolean functions of primary outputs under a heuristic ordering of input variables[MIY90]. We then transform the BDDs into prime-irredundant implicit cube sets by using the ISOP algorithm. The cube sets are then factorized into optimized multi-level logics by using the fast weak-division method. We wrote the program with C++ language on a SPARC station 2 (128 MB).

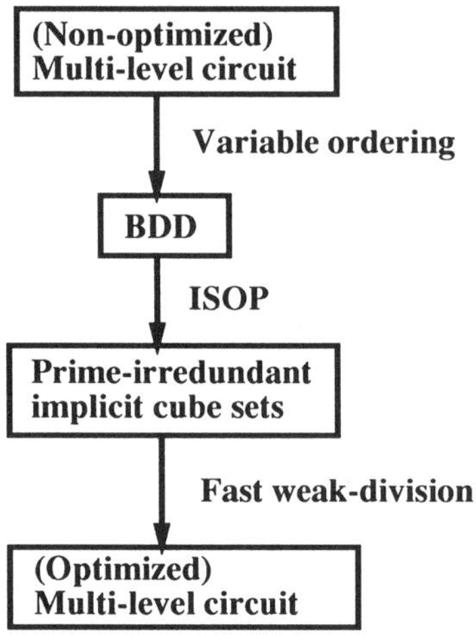

Figure 7.5 Basic flow of multi-level logic synthesizer.

We compared our experimental results with those of the MIS-II using a conventional cube-based method. The results are listed in Table 7.3. The circuits were an 8-bit and 16-bit parity functions "xor8" and "xor16," a (16+16)-bit adder "add16," (6×6)-bit multiplier "mult6," and other circuits chosen from the MCNC'90 benchmarks. The column "Lit." lists the total number of literals in multi-level logic, and the column "Time(s)" is the total CPU time for optimization. The heading "MIS-II(flatten)" labels the results obtained when we forced the MIS-II to flatten multi-level logics into cube sets and then factorized them using weak-division. "Our method" uses the implicit cube set representation.

The experimental results show that we can now quickly flatten and factorize circuits, even for parity functions and adders, that have never been flattened with other methods. Our method is much faster than MIS-II, and the difference is remarkable when large cube sets are factorized. This result demonstrates the effect of our implicit cube set representation.

When the flattened cube sets becomes too large, an incremental optimization method is sometimes effective. MIS-II provides an option of incremental optimization without flattening. We compared our results with such incremental methods. In Table 7.4, "MIS-II(algebra)" denotes the results of incremental optimization using only algebraic rules, and "MIS-II(Bool)" denotes the results obtained by using Boolean optimization

Table 7.3 Results of multi-level logic synthesis.

Func.	Two-Level Logic		MIS-II(flatten)		Our Method	
	Literals	ZBDD nodes	Literals	Time (sec)	Literals	Time (sec)
xor8	1,152	28	28	38.3	28	0.3
xor16	557,056	60	×	(>20h)	60	0.7
add16	11,468,595	176	×	(>20h)	257	6.9
mult6	22,273	3,315	×	(>20h)	6,802	2,900.7
9sym	1,036	42	83	29.8	117	1.8
vg2	914	102	97	33.9	102	1.7
alu4	5,539	1,129	1,319	3,751.6	1,148	64.5
apex1	4,115	1,768	2,863	10,945.1	2,521	209.6
apex2	15,530	1,144	×	(>20h)	253	29.5
apex3	4,679	1,539	2,132	1,926.6	2,221	158.2
apex4	8,055	1,545	3,509	1,345.9	3,473	462.4
apex5	7,603	2,387	1,206	156.9	1,185	58.7
c432	969,028	14,407	×	(>20h)	1,510	692.3

based on BDDs. Our method is inferior to these methods in terms of optimization quality (for example, on "mult6" and "alu4"), but incremental methods greatly depend on the characteristics of the initial circuits. With "9sym" and "vg2," for example, they are trapped in a local optimum. The results show that our method can be a good alternative method for logic optimization. In terms of computation time, our method is much faster than "MIS-II(Bool)," and is almost as fast as "MIS-II(algebra)."

7.4 CONCLUSION

We have developed a fast weak-division method for implicit cube sets based on ZBDDs. Computation time for this method is nearly proportional to the size of ZBDDs, and is independent of the number of cubes and literals in cube sets. Experimental results indicate that we can quickly flatten and factorize circuits — even for parity functions and adders, which have never been practicable before. Our method greatly accelerates logic synthesis systems and enlarges the scale of the circuits to which these systems are applicable.

Table 7.4 Comparison with incremental optimization.

Func.	Initial Literals	MIS-II(algebra) Literals	MIS-II(algebra) Time (sec)	MIS-II(Bool) Literals	MIS-II(Bool) Time (sec)	Our Method Literals	Our Method Time (sec)
xor8	63	28	0.4	28	1.1	28	0.3
xor16	135	60	0.9	60	1.9	60	0.7
add16	432	208	4.7	192	12.0	257	6.9
mult6	883	520	12.2	393	42.4	6,802	2,900.7
9sym	610	425	3.2	510	444.4	117	1.8
vg2	922	414	5.7	480	97.4	102	1.7
alu4	7,483	1,195	171.8	283	2,592.1	1,148	64.5
apex1	9,133	1,684	127.3	1,689	4,438.3	2,521	209.6
apex2	14,871	363	306.7	347	12,842.7	253	29.5
apex3	8,397	1,747	102.4	1,723	63,427.6	2,221	158.2
apex4	14,960	2,586	238.3	2,514	9,160.2	3,473	462.4
apex5	7,106	917	59.3	822	267.0	1,185	58.7
c432	372	265	4.4	331	728.9	1,510	692.3

There is still some room to improve the results. We have used a simple strategy for choosing divisors, but more sophisticated strategies might be possible. Moreover, a Boolean division method for implicit cube sets is worth investigating to improve the optimization quality.

8

IMPLICIT MANIPULATION OF POLYNOMIALS BASED ON ZBDDS

In this chapter, we present a good application of ZBDDs, that manipulates arithmetic polynomial formulas containing higher-degree variables and integer coefficients. In this method[Min95], we can represent large-scale polynomials compactly and uniquely, and can manipulate them in a practical time. Constructing canonical forms of polynomials immediately leads to equivalence checking of arithmetic expressions. Since polynomial calculus is a basic part of mathematics, our method is expected to be useful for various problems.

8.1 REPRESENTATION OF POLYNOMIALS

Polynomials can be manipulated in a similar way to the cube sets manipulation, except that polynomials differ from cube sets in the following two points:

- They can have higher-degree variables.
  ($x \cdot x = x^2$, rather than $x \cdot x = x$ in Boolean algebra.)
- They can have terms with coefficients.
  ($x + x = 2x$, rather than $x + x = x$ in Boolean algebra.)

Here we show a method for manipulating polynomials with higher-degree variables and coefficients.

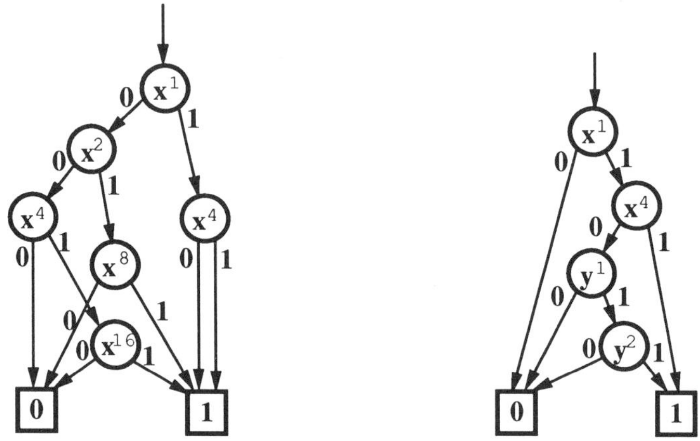

(a) ZBDD for $(x^4x^{16}+x^2x^8+x^1x^4+x^1)$.

(b) ZBDD for $(x^1x^4 + x^1y^1y^2)$.

Figure 8.1 Representation of higher-degrees.

8.1.1 Representation of Higher-Degrees

First, we show a way to deal with degrees. Here, we consider only degrees of positive integers. The basic idea is that an integer number can be written as a sum of 2's exponential numbers by using binary coding. That is, a variable x^k can be broken down as follows:

$$x^k = x^{(k_1+k_2+\ldots+k_m)} = x^{k_1}x^{k_2}\ldots x^{k_m},$$

where $k_1, k_2, \ldots, k_m$ are different 2's exponential numbers. Thus we can represent x^k as a combination of n items $x^1, x^2, x^4, x^8, \cdots, x^{2^{n-1}} (0 < k < 2^n)$, and we can efficiently deal with such combinations by using ZBDDs. A polynomial $x^{20} + x^{10} + x^5 + x$, for example, can be written as $x^4x^{16} + x^2x^8 + x^1x^4 + x^1$. It can be regarded as a *set of combinations* based on the five items x^1, x^2, x^4, x^8, and x^{16}. The formula can then be represented by using a ZBDD, as shown in Fig. 8.1(a). In this example, we ordered x^1 the top and ordered the higher degrees lower in the graph. This ordering is convenient in calculating arithmetic operations, which is described in Section 8.2.

When dealing with more than one sort of variable — such as x^i, y^j, and z^k — we decompose them as $x^1, x^2, x^4, \ldots$, $y^1, y^2, y^4, \ldots$, and $z^1, z^2, z^4, \ldots$. Figure 8.1(b) shows an example with two sorts of variables. Since our BDD package allows 65,535 variables, we can use more than 8,000 sorts of variables when using 8-bit coding (max 255) for degrees.

One of feature of our method is that it gives canonical forms of a polynomials, since the degrees are uniquely decomposed into the combinations based on a binary coding, and ZBDDs represent the sets of combinations uniquely. In addition, ZBDDs clearly exhibit their efficiency. For example, $x^1x^2(=x^3)$ is represented as a combination of only x^1 and x^2, but $x^4, x^8, x^{16}, \ldots$ are not included. In ZBDDs, the nodes for the irrelevant items $x^4, x^8, x^{16}, \ldots$ are automatically eliminated. In many cases, variables with lower degrees appear more often than those with higher degrees, so that most of the combinations are sparse and ZBDDs are effective. In addition, when dealing with many sorts of variables, we should consider that the combination x^1x^2 does not include other sorts of variables, such as $y^1, y^2, y^4, \ldots$, or $z^1, z^2, z^4, \ldots$. In this case, the combinations become very sparse and ZBDDs are especially effective.

8.1.2 Representation of Coefficients

Next, we present a way to represent coefficients. Here we consider only integer numbers as coefficients. The fundamental constants "0" and "1" are represented by 0- and 1-terminal nodes in ZBDDs. Another constant number c (> 1) can be written using binary coding as a sum of 2's exponential numbers:

$$c = 2^{c_1} + 2^{c_2} + \ldots + 2^{c_m},$$

where $c_1, c_2, \ldots, c_m$ are different positive integer numbers. If we regards "2" as a symbol, just like x, y, z, etc., we can represent it as a polynomial of variables with degrees, which has already been discussed. Consequently, we can represent a constant number c using ZBDDs as a set of combinations from n items $2^1, 2^2, 2^4, 2^8, \ldots, 2^{2^{n-1}}$ $(0 < c < 2^{2^n})$. For example, the constant $300 = 2^8 + 2^5 + 2^3 + 2^2$ can be written as $2^8 + 2^1 2^4 + 2^1 2^2 + 2^2$. It can be regarded as a set of combinations based on four items $2^1, 2^2, 2^4$, and 2^8, then represented by a ZBDD, as shown in Fig. 8.2(a).

When a constant number is used as a coefficient with other variables, we can regard the symbol "2" just as one sort of variable in the formula. Figure 8.2(b) shows an example for $5x^2 + 2xy$, which is decomposed into $x^2 + 2^2x^2 + 2^1x^1y^1$.

When dealing with negative coefficients, we have to consider the coding of negative values. There are two well-known methods, one using 2's complement representation, and the other using the absolute value with sign. Both methods, however, have drawbacks. The one using 2's complement yields many non-zero bits for small

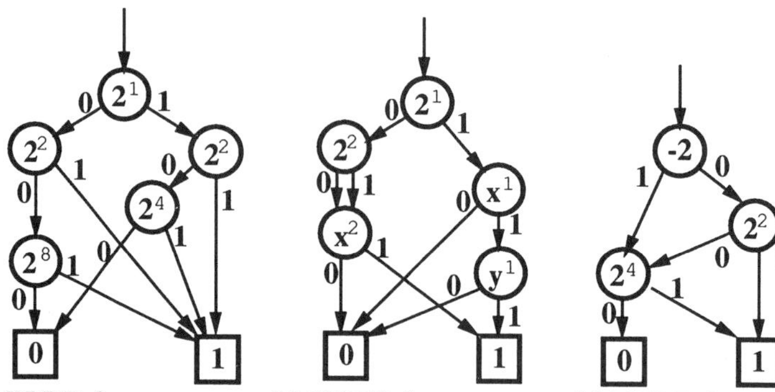

(a) ZBDD for $(2^8 + 2^1 2^4 + 2^1 2^2 + 2^2)$. (b) ZBDD for $(x^2 + 2^2 x^2 + 2^1 x^1 y^1)$. (c) ZBDD for $(-2 \cdot 2^4 + 2^4 + 2^2)$.

Figure 8.2 Representation of coefficients.

negative numbers (typically, -1 is "all one"), and the ZBDD reduction rule is not effective to those non-zero bits. And with the other using the absolute value, the operation of addition become complicated since we have to switch the addition into subtraction for some product terms in the same formula.

To solve these problems, we used another binary coding based on $(-2)^k$: each bit represents $1, -2, 4, -8, 16, \ldots$. For example, -12 can be decomposed into $(-2)^5 + (-2)^4 + (-2)^2 = -2 \cdot 2^4 + 2^4 + 2^2$, and can be represented by a ZBDD as shown in Fig. 8.2(c). In this way, we can avoid producing many non-zero bits for small negative numbers.

Two polynomials are equivalent if and only if they have the same coefficients for all corresponding terms. Since our new representation method maintains uniqueness, we can immediately check the equivalence between two polynomials after generating ZBDDs.

8.2 ALGORITHMS FOR ARITHMETIC OPERATIONS

Polynomials can be manipulated by arithmetic operations, such as addition, subtraction, and multiplication. We first generate ZBDDs for trivial formulas that are single variables or constants, and then we use these arithmetic operations to construct more complicated polynomials. An example is shown in Fig. 8.3. To generate a ZBDD

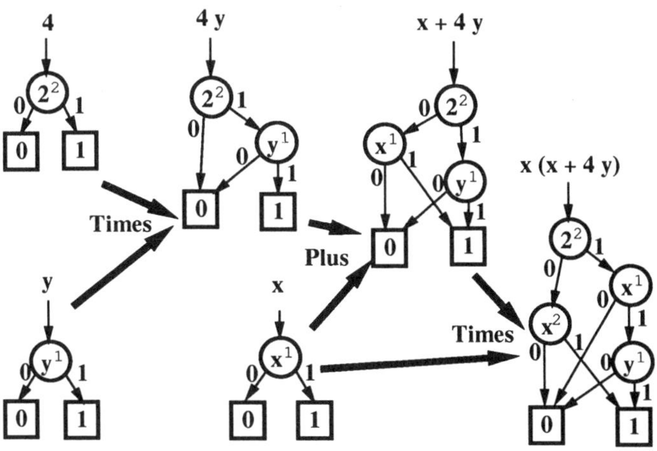

Figure 8.3 Generation of ZBDDs for arithmetic expressions.

for the formula $x^2 + 4xy$ from the arithmetic expression $x \times (x + 4 \times y)$, we first generate graphs for "x," "y," and "4," and then use some arithmetic operations. After generating ZBDDs for polynomials, we can immediately check the equivalence between two polynomials. We can also easily evaluate the polynomials in terms of the length, degrees, coefficients, etc.

In this section, we present efficient algorithms for the arithmetic operations of polynomials.

8.2.1 Multiplication by a Variable

We first show an algorithm for multiplying a polynomial F by a variable v. This operation is a basic part of other arithmetic operations. The algorithm divides F into the two subformulas F_1 and F_0 according to whether they contain v. In multiplying by v, each product term in F_0 gets v, and each product term in F_1 gets v^2 instead of v. Then $(F_1 \times v^2)$ is computed recursively. This action can be described as
$F \times v = v \cdot F_0 \cup (F_1 \times v^2)$, where $F = v \cdot F_1 \cup F_0$,
and is illustrated in Fig. 8.4. The algorithm is executed efficiently when the variables are ordered as $x^1, x^2, x^4, x^8, \ldots$; that is, $(x^k)^2$ is always the next variable of x^k.

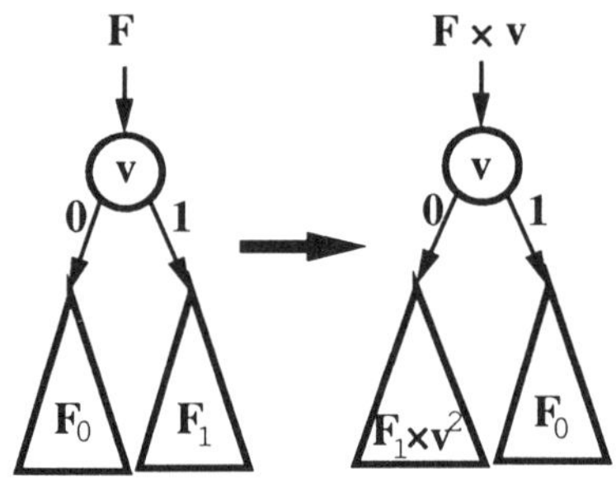

Figure 8.4 ZBDDs in multiplication by a variable.

By multiplying by a special symbol 2^k (or $(-2)^k$), we can perform a "shift" operation for the coefficients in a formula.

8.2.2 Addition

If F and G have no common combinations, the addition $(F + G)$ can be completed by just merging them. When they contain some common combinations, we calculate the following formula:

$(F + G) = S + (C \times 2)$,

where $C = F \cap G,\ S = (F \cup G) - C$.

By repeating this process, we eventually exhaust common combinations and complete the procedure.

We can explain the action of the algorithm using an example: the addition of $F = x + z$ and $G = 3x + y\ (= 2^1x + x + y)$. In the first execution, $C \leftarrow x$ and $S \leftarrow 2^1x + y + z$. Since $C \neq 0$, we repeat the procedure with $F = 2^1x + y + z\ (= S)$ and $G = 2^1x\ (= C \times 2)$. In the second execution, $C \leftarrow 2^1x$ and $S \leftarrow y + z$, and we repeat with $F = y + z$ and $G = 2^2x$. The third times, $C = 0$ and the result $2^2x + y + z$ is obtained.

When the coding based on $(-2)^k$ is used, the addition and subtraction can be performed as follows:

$(F + G) = S - (C \times (-2))$,

$(F - G) = D + (B \times (-2))$,

where $D = F \cap \overline{G},\ B = \overline{F} \cap G.$
In this procedure, the addition and subtraction are called alternately.

8.2.3 Multiplication of Polynomials

Using the two operations described in previous section, we can compose an algorithm for multiplying two polynomials. This algorithm is based on the divide-and-conquer idea. Suppose that v is the variable of the root node in the ZBDD. We first factor F and G into two parts:

$$F = v \cdot F_1 \cup F_0, \quad G = v \cdot G_1 \cup G_0.$$

The product $(F \times G)$ can then be written as:

$$\begin{aligned}(F \times G) \quad &= \quad (F_0 \times G_0) + (F_1 \times G_1) \times v^2 \\ &\quad\quad + ((F_1 \times G_0) + (F_0 \times G_1)) \times v.\end{aligned}$$

Each sub-product term can be computed recursively, and the result is obtained when the expressions are eventually broken down into trivial ones. In the worst case, this algorithm would require a number of recursive calls that is exponential for the number of variables; however, we can accelerate the procedure by using a hash-based cache memory that stores the results of recent operations. By referring to the cache before each recursive call, we can avoid duplicate executions for equivalent subformulas. Consequently, the execution time depends on the size of the ZBDDs, not on the number of terms. This algorithm is given in detail in Fig. 8.5.

8.2.4 Division

In polynomial calculus, the division operation yields a quotient (F/G) and a remainder $(F\%G)$. This algorithm can also be described by a recursive formula. Suppose x is the variable of the root node in ZBDD, and that s and t are the highest degrees of x in F and G, respectively. They are then factored into two parts:

$$F = x^s \cdot F_1 \cup F_0, \quad G = x^t \cdot G_1 \cup G_0.$$

The quotient (F/G) can be written as:

$$(F/G) \quad = \quad (F_1/G_1)x^{(s-t)} + (F - (F_1/G_1)x^{(s-t)} \times G)/G.$$

The sub-quotient terms can be computed recursively. As in the multiplication algorithm, the result is obtained when the expressions are eventually broken down into

```
procedure(F × G)
{    if (F.top < G.top) return (G × F) ;
     if (G = 0) return 0 ;
     if (G = 1) return F ;
     H ← cache("F × G") ; if (H exists) return H ;
     v ← F.top ; /* the highest variable in F */
     (F0, F1) ← factors of F by v ;
     (G0, G1) ← factors of G by v ;
     H ← (F1 × G1) × v^2
        +((F1 × G0) + (F0 × G1)) × v + (F0 × G0) ;
     cache("F × G") ← H ;
     return H ;
}
```

Figure 8.5 Algorithm for multiplication of polynomials.

trivial ones. (For example, $F/G = 0$ when $s < t$.) A hash-based cache memory is also referred to avoid duplicate executions for equivalent subformulas. The remainder $(F\%G)$ can be obtained by $F - (F/G) \times G$.

When dealing with more than one sort of variables, this algorithm may give different quotients depending on the variable ordering of ZBDDs. However, the quotient is computed uniquely for any variable ordering when the remainder is zero. Using this division algorithm, we can easily check whether a polynomial is a factor of another polynomial.

8.2.5 Substitution

Using the division operation, we can compute the substitution operation $F[x = G]$ for given polynomials F and G, and for a variable x contained in F. The algorithm can be written as:

$$F[x = G] = (F/x)[x = G] \times G + (F\%x).$$

$(F/x)[x = G]$ is computed recursively. If no x appears in F, $F[x = G] = F$.

We can use this algorithm to perform various substitutions, such as $F[x = x{+}1]$, $F[x = y^2]$, and $F[x = 5]$. This operation is very useful for practical applications.

Table 8.1 Result for $n!$.

n	ZBDD nodes	Time (sec)
10	8	0.1
20	22	0.3
30	32	0.9
40	43	1.7
50	64	2.8
56	62	3.2

8.3 IMPLEMENTATION AND EXPERIMENT

Using the techniques described in Section 8.2, we implemented a program for manipulating polynomials. Our program is written in C++ language on a SPARC station 2 (SunOS 4.1.3, 32 MB). It can handle about 8,000 sorts of variables, degrees up to 255, and coefficients up to 2^{255}. Our BDD package requires about 30 bytes per node.

To evaluate our method, we constructed ZBDDs for large-scale polynomials. We first generated ZBDDs for constants. In our experiment, only 15 nodes were needed to represent the number "1,000,000,000." Table 8.1 lists the results for $n!$. We can easily generate ZBDDs for constants as large as $56!$ within about three seconds. (When $n = 57$, $n!$ exceeds 256 bit.)

We also tried to represent various kinds of polynomials, such as x^n, $(x + 1)^n$, $\sum_{k=0}^{n} x^k$, and $\prod_{k=1}^{n}(x_k + 1)$. As shown in Tables 8.2 and 8.3, within a feasible time and space, we can generate ZBDDs for extremely large-scale polynomials, some of which consist of millions of terms. This has never been practical in conventional representation, which requires a memory space proportional to the number of terms.

Our method greatly accelerates the computation of polynomials and expands the scale of their applicability. It is especially effective when dealing with many sorts of variables, a feat that has been difficult for conventional methods.

Table 8.2 Results for representing polynomials.

n	x^n		n^2x^n		$\sum_{k=0}^{n} x^k$		$\sum_{k=1}^{n}(k \times x^k)$		$(x+1)^n$	
	ZBDD nodes	Time (sec)	ZBDD nodes	Time (sec)	ZBDD nodes	Time (sec)	ZBDD nodes	Time (sec)	ZBDD nodes	Time (sec)
1	1	0.1	1	0.1	1	0.1	1	0.1	1	0.1
2	1	0.1	2	0.1	2	0.1	4	0.1	4	0.1
3	2	0.1	5	0.1	2	0.1	5	0.1	5	0.1
5	2	0.1	6	0.1	3	0.1	8	0.1	10	0.1
10	2	0.1	7	0.1	6	0.1	18	0.1	23	0.2
20	2	0.1	7	0.1	8	0.1	26	0.1	69	0.5
30	4	0.1	9	0.1	8	0.1	37	0.2	150	2.0
50	3	0.1	11	0.1	10	0.1	48	0.3	346	7.1
100	3	0.1	11	0.1	12	0.1	70	0.5	1,209	39.7
200	3	0.1	12	0.1	14	0.1	84	1.0	4,231	267.7
255	8	0.1	11	0.1	8	0.2	75	1.3	6,690	528.8

Table 8.3 Results for large-scale polynomials.

n	$\prod_{k=1}^{n}(x_k+1)$			$\prod_{k=1}^{n}(x_k+k)$			$\prod_{k=1}^{n}(x_k+1)^4$		
	Terms	ZBDD nodes	Time (sec)	Terms	ZBDD nodes	Time (sec)	Terms	ZBDD nodes	Time (sec)
1	2	1	0.1	2	1	0.1	5	7	0.1
2	4	2	0.1	4	4	0.1	25	25	0.1
3	8	3	0.1	8	10	0.1	125	70	0.1
5	32	5	0.1	32	32	0.1	3,125	264	0.3
10	1,024	10	0.1	1,024	619	1.1	9,765,625	2,053	2.8
11	2,048	11	0.1	2,048	1,131	3.5	48,828,125	2,730	4.9
12	4,096	12	0.1	4,096	1,866	4.8	244,140,625	3,575	6.7
13	8,192	13	0.1	8,192	3,334	8.4	1,220,703,125	4,586	8.8
15	32,768	15	0.1	32,768	9,338	22.8	(5^{15})	7,151	19.7
20	1,048,576	20	0.1	–	(*)	–	(5^{20})	17,374	78.5

(*) Memory overflow (32 MB).

8.4 APPLICATION FOR LSI CAD

Because polynomial calculus is a basic part of mathematics, our method is useful for various problems in LSI CAD. One good application is in the computing of signal

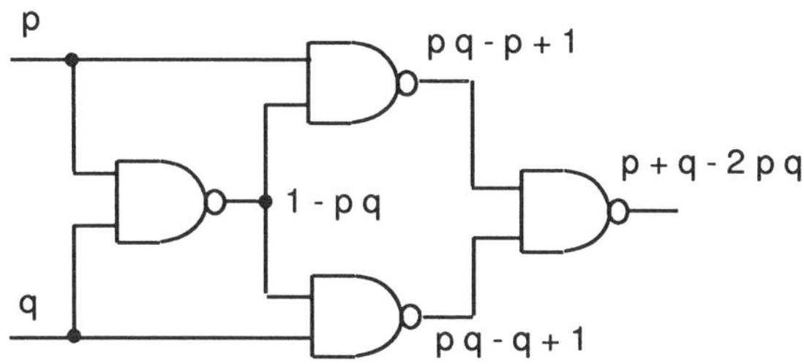

Figure 8.6 Computation of signal probability in logic circuit.

probability in logic circuits. As illustrated in Fig. 8.6, on each primary input of the circuit, we assign a variable representing the probability that the signal is '1.' Then the probability at primary outputs and internal nets can be expressed exactly by polynomials using those probabilistic variables. These polynomials may grow large, but our method nonetheless helps us to manipulate them efficiently. In our preliminary experiment, the polynomial for an 8-bit adder circuit required as many as 9,841 product terms, but it could be represented by a ZBDD with only 125 nodes. This technique is applicable for various kinds of statistic analysis, such as probabilistic fault simulation, power consumption estimation, and timing hazard analysis.

The formal verification of arithmetic-level description is another possible application. For example, suppose the two arithmetic expressions:

$F_1 = (z - 2x)(6xy - 15xz + 2y^2 - 5yz)$,
$F_2 = (3x + y)(10xz - 4xy + 2yz - 5z^2)$,
and whether $(F_1 = F_2)$?

By generating canonical forms of polynomials for the two expressions, the equivalence can be checked immediately. In addition, we can easily calculate the difference between the two expressions, for finding design error when they are not equivalent. Our method would thus be useful as a basic technique for high-level synthesis and formal verification in LSI design.

8.5 CONCLUSION AND REMARKS

We have proposed an elegant way to represent polynomials by using ZBDDs, and have shown efficient algorithms for manipulating those polynomials. Our experimental results indicate that we can manipulate large-scale polynomials implicitly within a

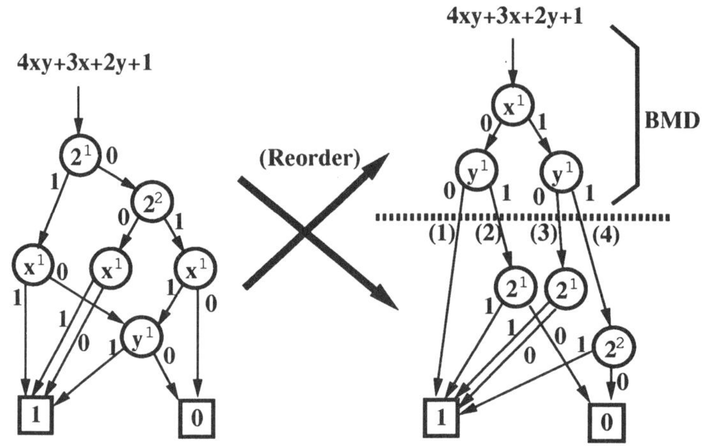

(a) Coefficients are higher. (b) Coefficients are lower.

Figure 8.7 Comparison of ZBDD-based method and BMD.

feasible time and space. An important feature of our representation is that it is the canonical form of a polynomial under a fixed variable order. Because the polynomial calculus is a basic part of mathematics, our method is very useful in various areas.

Recently, several variants of BDDs have been devised to represent multi-valued logic functions, and one of the remarkable idea is *Binary Moment Diagram (BMDs)* developed by Bryant[BC95] and mentioned in Section 4.3. BMDs can also represent polynomials containing coefficients. One big difference is that BMDs assume binary-valued variables, so they cannot deal with the higher-degree variables. Except for this difference, the BMDs and our method are near to each other. Figure 8.7(a) shows the ZBDD-based representation for $(4xy + 3x + 2y + 1)$. If we change the variable order such that the coefficient variables move from higher to lower positions, the ZBDD becomes as shown in Fig. 8.7(b), where the subgraphs with the coefficient variables correspond to the terminal nodes in the BMD. This indicates that the BMD and the ZBDD-based representation can be transformed into each other by changing the variable order. The efficiency of the two representations therefore depends on the nature of the objective functions; it is thus difficult to determine which representation is more efficient in general.

In the future we will try to implement other interesting operations on polynomials, such as differential methods, finding max/min values, solving equations, approximation, and factorization. We can generalize this polynomial manipulation technique to include such things as rational expressions, negative or non-integer degrees, and complex(imaginary)-number coefficients.

9

ARITHMETIC BOOLEAN EXPRESSIONS

9.1 INTRODUCTION

Boolean expressions are sometimes used in the research and development of digital systems, but calculating Boolean expressions by hand is a cumbersome job , even when they have only a few variables. For example, the Boolean expressions $(a\wedge\overline{b})\vee(\overline{a}\wedge\overline{c})\vee(b\wedge c)$, $(a\wedge\overline{b})\vee(a\wedge c)\vee(\overline{a}\wedge b)\vee(\overline{a}\wedge\overline{c})$, and $(a\wedge\overline{b})\vee(\overline{a}\wedge b)\vee(b\wedge c)\vee(\overline{b}\wedge\overline{c})$ represent the same function, but it is hard to verify their equivalence by hand. If they have more than five or six variables, we might as well give up. This problem motivated us to develop a Boolean expression manipulator (BEM)[MIY89], which is an interpreter that uses BDDs to calculate Boolean expressions. It enables us to check the equivalence and implications of Boolean expressions easily, and it helped us in developing VLSI design systems and solving combinatorial problems.

Most discrete problems can be described by logic expressions; however, the arithmetic operators such as addition, subtraction, multiplication, and comparison, are convenient for describing many practical problems. Such expressions can be rewritten using logic operators only, but this can result in expressions that are complicated and hard to read. In many cases, arithmetic operators provide simple descriptions of problems.

In this chapter, we present a new Boolean expression manipulator, that allows the use of arithmetic operators[Min93a]. This manipulator, called BEM-II, can directly solve problems represented by a set of equalities and inequalities, which are dealt with in 0-1 linear programming. Of course, it can also manipulate ordinary Boolean expressions. We developed several output formats for displaying expressions containing arithmetic operators.

In the following sections, we first show a method for manipulating Boolean expressions with arithmetic operations. We then present implementation of the BEM-II and its applications.

9.2 MANIPULATION OF ARITHMETIC BOOLEAN EXPRESSIONS

Most discrete problems can be described by using logic operations; however, arithmetic operators are useful for describing many practical problems. For example, a majority function with five inputs can be expressed concisely by using arithmetic operators:

$$x_1 + x_2 + x_3 + x_4 + x_5 \geq 3.$$

When, on the other hand, it is written using only Boolean expressions, this function becomes more complicated:

$$\begin{array}{l}(x_1 \wedge x_2 \wedge x_3) \vee (x_1 \wedge x_2 \wedge x_4) \vee (x_1 \wedge x_2 \wedge x_5) \\ \vee (x_1 \wedge x_3 \wedge x_4) \vee (x_1 \wedge x_3 \wedge x_5) \vee (x_1 \wedge x_4 \wedge x_5) \\ \vee (x_2 \wedge x_3 \wedge x_4) \vee (x_2 \wedge x_3 \wedge x_5) \vee (x_2 \wedge x_4 \wedge x_5) \\ \vee (x_3 \wedge x_4 \wedge x_5).\end{array}$$

In this section, we describe an efficient method for representing and manipulating Boolean expressions containing arithmetic operators.

9.2.1 Definitions

We define *arithmetic Boolean expressions* and *Boolean-to-integer functions*, which are extended models of conventional Boolean expressions and Boolean functions.

Definition 9.1 Arithmetic Boolean expressions *are extended Boolean expressions that contain not only logic operators, but also arithmetic operators, such as addition* $(+)$*, subtraction* $(-)$*, and multiplication* $(\times)$*. Any integer can be used as a constant in the expression, but input variables are restricted to either 0 or 1. Equality* $(=)$ *and inequalities* $(<, >, \leq, \geq, \neq)$ *are defined as operations returning a value of either 1 (true) or 0 (false).* □

For example, $(3 \times x_1 + x_2)$ is an arithmetic Boolean expression with respect to the variables $x_1, x_2 \in \{0, 1\}$. $(3 \times x_1 + x_2 < 4)$ is also an example.

	x_1x_2			
	00	01	10	11
$3 \times x_1$	0	0	3	3
$3 \times x_1 + x_2$	0	1	3	4
$3 \times x_1 + x_2 < 4$	1	1	1	0

Figure 9.1 Computation of arithmetic Boolean expressions.

When ordinary logic operations are applied to integer values other than 0 and 1, we define them as bit-wise logic operations for binary-coded numbers, like the manner in many programming languages. For example, $(3 \vee 5)$ returns 7. Under this modeling scheme, conventional Boolean expressions become special cases of arithmetic Boolean functions.

The value of the expression $(3 \times x_1 + x_2)$ becomes 0 when $x_1 = x_2 = 0$, and it becomes 4 when $x_1 = x_2 = 1$. We can see that an arithmetic Boolean expression represents a function from a binary-vector to an integer: $(B^n \to I)$. We call this a **Boolean-to-integer (B-to-I) function**, which has been discussed in Section 4.2. The operators in arithmetic Boolean expressions are defined as operations on B-to-I functions.

The procedure for obtaining the B-to-I function for the arithmetic Boolean expression $(3 \times x_1 + x_2 < 4)$ is shown in Fig. 9.1. First, multiply the constant function 3 by the input function x_1 to obtain the B-to-I function for $(3 \times x_1)$. Then add x_2 to obtain the function for $(3 \times x_1 + x_2)$. Finally we can get a B-to-I function for the entire expression $(3 \times x_1 + x_2 < 4)$ by applying the comparison operator ($<$) to the constant function 4. We find that this arithmetic Boolean expression is equivalent to the expression $(\overline{x_1} \vee \overline{x_2})$.

9.2.2 Representation of B-to-I Functions

As shown in Fig. 9.1, a B-to-I function can be obtained by enumerating the output values for all possible combinations of the input values. But because this procedure is impracticable when there are many input variables (since the number of combinations grows exponentially), we need a more efficient way to represent B-to-I functions.

As discussed in Section 4.2, there are two ways to represent B-to-I functions: multi-terminal BDDs (MTBDDs) and BDD vectors. MTBDDs are extended BDDs with

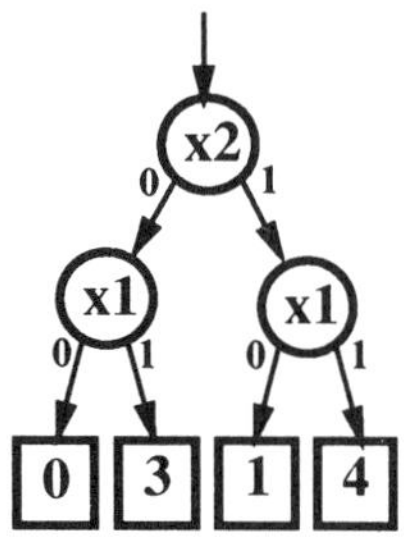

(a) A MTBDD.

B-to-I function

x1	x2	f (f2 f1 f0)
0	0	0 (0 0 0)
0	1	1 (0 0 1)
1	0	3 (0 1 1)
1	1	4 (1 0 0)

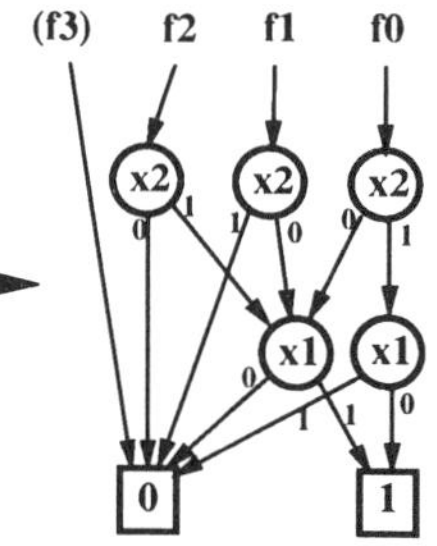

(b) A BDD vector.

Figure 9.2 Representation for B-to-I functions.

multiple terminals, each of which has an integer value (Fig. 9.2(a)), and BDD vectors are the way to represent B-to-I functions with a number of BDDs by encoding the integer numbers into binary vectors (Fig. 9.2(b)).

The efficiency of the two representations depends on the nature of the objective functions. In manipulating arithmetic Boolean expressions, we often generate B-to-I functions from Boolean functions, as when we calculate $F \times 2$, $F \times 5$, or $F \times 100$ from a certain Boolean function F. In such cases, the BDD vectors can be conveniently shared with each other (Fig. 9.3). Multi-terminal BDDs, however, cannot be shared (Fig. 9.4), so we use BDD vectors for manipulating arithmetic Boolean expressions.

For negative numbers, we use 2's complement representation in our implementation. The most significant bit is used for the sign bit, whose BDD indicates the condition under which the B-to-I function produces a negative value. This coding scheme requires that the word-length be specified in order to know which is sign bit. An easy way is to allocate a long length in advance, but this limits the range of numbers. Our implementation supports a variable word-length for each B-to-I function, so there is no limit on the range of numbers.

9.2.3 Handling B-to-I functions

This section explains how to handle B-to-I functions represented by BDD vectors. Logic operations — such as AND, OR, and EXOR — are implemented as bit-wise operations between two BDD vectors. Applying BDD operations to their respective bits generates a new B-to-I function. We define two kinds of inversion operations:

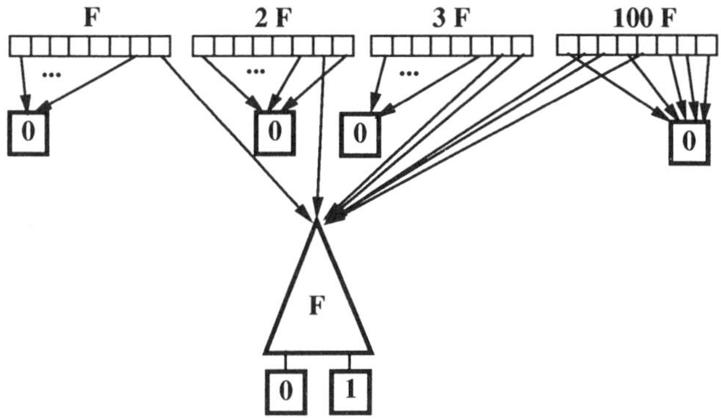

Figure 9.3 BDD vectors for arithmetic Boolean expressions.

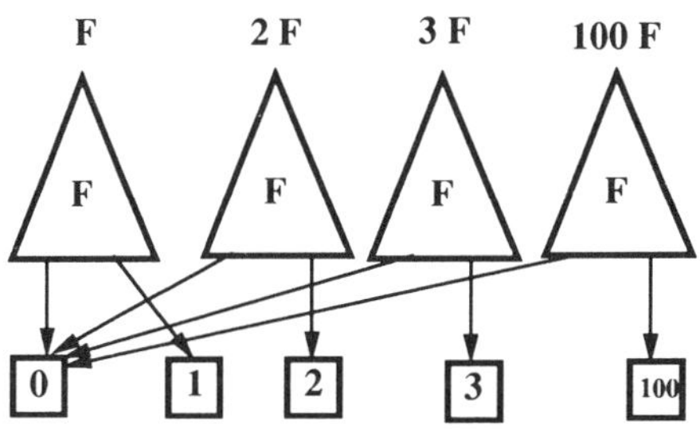

Figure 9.4 MTBDDs for arithmetic Boolean expressions.

bit-wise inversion and logical inversion. Logical inversion returns 1 for 0, and returns 0 for the other numbers.

Arithmetic addition can be composed using logic operations on BDDs by simulating a conventional hardware algorithm of full-adders designed as combinational circuits. We use a simple algorithm for a ripple carry adder, which computes from the lower bit to the higher bit, propagating carries. Other arithmetic operations like subtraction, multiplication, division, and shifting can be composed in the same way. Exception handling should be considered for overflow and division by zero.

Positive/negative checking is immediately performed by returning the sign-bit BDD. Using subtraction and sign checking, we can compose comparison operations between two B-to-I functions. These operations generate a new B-to-I function that returns a value of either 1 or 0 to express whether the equality or inequality is satisfied.

It would be useful to find the upper (or lower) bound value of a B-to-I function for all possible combinations of input values. This can be done efficiently by using a binary search. To find the upper bound, we first check whether the function can ever exceed 2^n. If there is a case in which it does, we then compare it with $2^n + 2^{n-1}$, otherwise only with 2^{n-1}. In this way, all the bits can be determined from the highest to the lowest, and eventually the upper bound is obtained. The lower bound is found in a similar way.

Computing the upper (or lower) bound is a unary operation for B-to-I functions; it returns a constant B-to-I function and can be used conveniently in arithmetic Boolean expressions. For example, the expression:

$$UpperBound(F) = F \quad (F \text{ is an arithmetic Boolean expression})$$

gives a function that returns 1 for the inputs that maximize F, otherwise it returns 0. That is, it computes the condition for maximizing F.

An example of calculating arithmetic Boolean expressions using BDD vectors is shown in Fig. 9.5.

9.2.4 Display Formats for B-to-I Functions

We propose several formats for displaying B-to-I functions represented by BDDs.

Integer Karnaugh Maps

A conventional Karnaugh map displays a Boolean function by using a matrix of logic values (0, 1). We extended the Karnaugh map to include integers as elements (Fig. 9.6). We call this *integer Karnaugh map*. It is useful for observing the behavior of B-to-I functions. Like ordinary Karnaugh maps, they are practical only for functions with fewer than five or six inputs. For a larger number of inputs, we can make an integer Karnaugh map with respect to only six input variables, by displaying the upper (or lower) bound for the rest of variables on each element of the map.

Bitwise Expressions

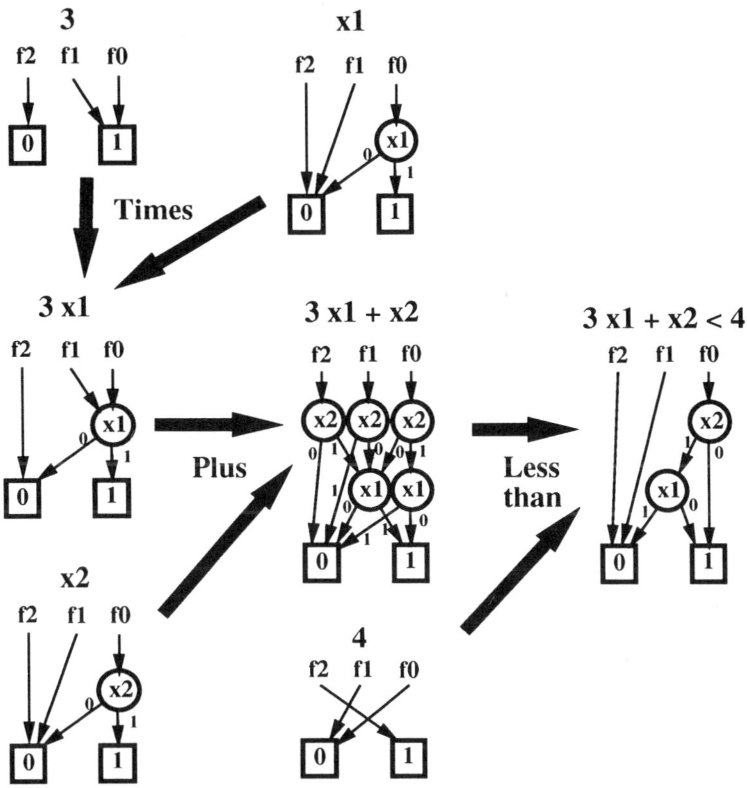

Figure 9.5 Generation of BDD vectors for arithmetic Boolean expressions.

When the objective function is too complicated for an integer Karnaugh map, the function can be displayed by listing Boolean expressions for respective bits of the BDD vector in the sum-of-products format. Figure 9.7 shows a bitwise expression for the same function shown in Fig .9.6. We used the ISOP algorithm (described in Chapter 5) to generate a compact sum-of-products format on each bit.

Bitwise expression is not very helpful for showing the behavior of B-to-I functions, but it does allow us to observe the appearance frequency of an input variable and it can estimate a kind of complexity of the functions.

F = 2 a + 3 b - 4 c + d

ab \ cd	00	01	11	10
00	0	1	-3	-4
01	3	4	0	-1
11	5	6	2	1
10	2	3	-1	-2

Figure 9.6 An integer Karnaugh map.

If a function never has negative values, we can suppress the expression for the sign bit. If some higher bits are always zero, we can omit showing them. With this zero suppression, a bitwise expression becomes a simple Boolean expression if the function returns only 1 or 0.

Case Enumeration

Using case enumeration, we can list all possible values of a function and display the condition for each case using a sum-of-products format (Fig. 9.8). This format is effective when there are many input variables but the range of output values is limited.

Arithmetic Sum-of-Products Format

It would be useful if we could display a B-to-I function as an expression using arithmetic operators. There is a trivial way of generating such an expression by using the case enumeration format. When the case enumeration method gives the values $v_1, v_2, \ldots, v_m$ and their conditions $F_1, F_2, \ldots, F_m$, we can create the expression $(v_1 \times F_1 + v_2 \times F_2 + \ldots + v_n \times F_m)$. In this method, the expression $(2 \times a + 3 \times b - 4 \times c + d)$ would thus be displayed as

$$
\begin{aligned}
&6 \times a\, b\, \overline{c}\, d + 5 \times a\, b\, \overline{c}\, \overline{d} + 4 \times \overline{a}\, b\, \overline{c}\, d + 3 \times (a\, \overline{b}\, \overline{c}\, d + \overline{a}\, b\, \overline{c}\, \overline{d}) \\
&+ 2 \times (a\, b\, c\, d + a\, \overline{b}\, \overline{c}\, \overline{d}) + (a\, b\, c\, \overline{d} + \overline{a}\, \overline{b}\, \overline{c}\, d) - (a\, \overline{b}\, c\, d + \overline{a}\, b\, c\, \overline{d}) \\
&- 2 \times a\, \overline{b}\, c\, d - 3 \times \overline{a}\, \overline{b}\, c\, d - 4 \times \overline{a}\, \overline{b}\, c\, \overline{d}.
\end{aligned}
$$

$$F = 2 \times a + 3 \times b - 4 \times c + d$$

$$\begin{array}{rcl}
\pm & : & (\overline{a} \wedge c \wedge \overline{d}) \vee (\overline{b} \wedge c) \\
F_2 & : & (a \wedge b \wedge \overline{c}) \vee (\overline{a} \wedge c \wedge \overline{d}) \vee (b \wedge \overline{c} \wedge d) \vee (\overline{b} \wedge c) \\
F_1 & : & (a \wedge \overline{b}) \vee (a \wedge d) \vee (\overline{a} \wedge b \wedge \overline{d}) \\
F_0 & : & (b \wedge \overline{d}) \vee (\overline{b} \wedge d)
\end{array}$$

Figure 9.7 A bitwise expression

$$\begin{array}{rcl}
6 & : & a \wedge b \wedge \overline{c} \wedge d \\
5 & : & a \wedge b \wedge \overline{c} \wedge \overline{d} \\
4 & : & \overline{a} \wedge b \wedge \overline{c} \wedge d \\
3 & : & (a \wedge \overline{b} \wedge \overline{c} \wedge d) \vee (\overline{a} \wedge b \wedge \overline{c} \wedge \overline{d}) \\
2 & : & (a \wedge b \wedge c \wedge d) \vee (a \wedge \overline{b} \wedge \overline{c} \wedge \overline{d}) \\
1 & : & (a \wedge b \wedge c \wedge \overline{d}) \vee (\overline{a} \wedge \overline{b} \wedge \overline{c} \wedge d) \\
0 & : & (\overline{a} \wedge b \wedge c \wedge d) \vee (\overline{a} \wedge \overline{b} \wedge \overline{c} \wedge \overline{d}) \\
-1 & : & (a \wedge \overline{b} \wedge c \wedge d) \vee (\overline{a} \wedge b \wedge c \wedge \overline{d}) \\
-2 & : & a \wedge \overline{b} \wedge c \wedge \overline{d} \\
-3 & : & \overline{a} \wedge \overline{b} \wedge c \wedge d \\
-4 & : & \overline{a} \wedge \overline{b} \wedge c \wedge \overline{d}
\end{array}$$

Figure 9.8 Case enumeration format.

This expression seems much more complicated than the original one, which has a linear form. Here we propose a method for eliminating the negative literals from the above expression and making an arithmetic sum-of-products expression which consists only of arithmetic addition, subtraction, and multiplication operators. Our method is based on the following expansion:

$$\begin{aligned} F &= x \times F_1 + \overline{x} \times F_0 \\ &= x \times (F_1 - F_0) + F_0, \end{aligned}$$

where F is the objective function, and F_0 and F_1 are subfunctions obtained by assigning 0 and 1 to input variable x. By recursively applying this expansion to all the input variables, we can generate an arithmetic sum-of-products expression containing no negative literals. We can thereby extract a linear expression from a B-to-I function represented by a BDD vector. For example, we can generate a BDD vector for $2 \times (a + 3 \times b) - 4 \times (a + b)$, and it can be displayed in a reduced format as $(-2 \times a + 2 \times b)$.

The arithmetic sum-of-products format seems unsuitable for representing ordinary Boolean functions. For example, $(a \wedge \overline{b}) \vee (c \wedge \overline{d})$ becomes $-a\,b\,c\,d + a\,b\,c - a\,b + a\,c\,d - a\,c + a - c\,d + c$, which is more difficult to read than the original one.

9.3 APPLICATIONS

Using the techniques described above, we developed an arithmetic Boolean expression manipulator, called BEM-II. It is an interpreter with a lexical and syntax parser for calculating arithmetic Boolean expressions and displaying the results in various formats. This section gives the specifications of BEM-II and discusses some of their applications.

9.3.1 BEM-II Specification

BEM-II[1] has a C-shell-like interface, both for interactive execution from the keyboard and for batch jobs from a script file. It parses the script only from left to right; neither branches nor loop controls are supported. The available operators is listed in Table 9.1, and an execution example is shown in Fig. 9.9.

In BEM-II scripts, we can use two kind of variables: *input variables* and *register variables*. Input variables, denoted by strings starting with a lowercase letter, represent the inputs of the functions to be computed. They are assumed to have a value of either 1 or 0. Register variables, denoted by strings starting with an uppercase letter, are used to identify the memory to which a B-to-I function is to be saved temporarily. We can describe multi-level expressions using these two types of variables; for example,

```
F = a + b ; G = F + c.
```

Calculation results are displayed as expressions with input variables only, without using register variables. BEM-II allows 65,535 different input variables to be used, and there is no limit on the number of register variables.

BEM-II supports not only logical operators such as AND, OR, EXOR, and NOT, but also arithmetic operators such as addition, subtraction, multiplication, division, shifting, equality, inequality, and upper/lower bound. The syntax for expressions

[1]This program, which runs on a SPARC station, is a public domain software.
FTP server address: eda.kuee.kyoto-u.ac.jp (130.54.29.134) /pub/cad/BemII

Table 9.1 Operators in BEM-II.

(The upper operators are executed before the lower ones.)

<table>
<tr><td>( )</td></tr>
<tr><td>! (logical) ~ (bit-wise) + −(unary)</td></tr>
<tr><td>* / (quotient) % (remainder)</td></tr>
<tr><td>+ −(binary)</td></tr>
<tr><td><< >> (bit-wise shift)</td></tr>
<tr><td>< <= > >= == != (relation)</td></tr>
<tr><td>& (bit-wise AND)</td></tr>
<tr><td>^ (bit-wise EXOR)</td></tr>
<tr><td>| (bit-wise OR)</td></tr>
<tr><td>? : (if-then-else)</td></tr>
<tr><td><code>UpperBound( ) LowerBound( )</code></td></tr>
</table>

generally conforms to C language specifications. The expression `A ? B : C` means *if-then-else*, and is equivalent to

```
((A != 0) * B) + ((A == 0) * C).
```

BEM-II generates BDD vectors of B-to-I functions for given arithmetic Boolean expressions. Since BEM-II can generate huge BDDs with millions of nodes, limited only by memory size, we can manipulate large-scale and complicated expressions. It can, of course, calculate any expressions that used to be manipulated by hand. The results can be displayed in the various formats presented in earlier sections.

The input variables are assumed to have values that are either 1 or 0, but multi-valued variables are sometimes used in real problems. In such cases, we can use register variables to deal with multi-valued variables. For example, `X = x0 + 2*x1 + 4*x2 + 8*x3` represents a variable ranging from 0 to 15. In another way, `X = x1 + 2*x2 + 3*x3 + 4*x4` represents a variable ranging from 1 to 4, under the one-hot constraint `(x1 + x2 + x3 + x4 == 1)`.

BEM-II can be used for solving many kind of combinatorial problems. Using BEM-II, we can generate BDDs for constraint functions of combinatorial problems specified by arithmetic Boolean expressions. This enables us to solve 0-1 linear programming problems by handling equalities and inequalities directly, without coding complicated procedures in a programming language. BEM-II can also solve problems which are expressed by non-linear expressions. BEM-II features its customizability: we can

```
% bemII
***** Arithmetic Boolean Expression Manipulator (Ver. 4.2) *****
> symbol a b c d
> F = 2*a + 3*b - 4*c + d
> print /map F
 a b : c d
       |    00    01    11    10
  00   |     0     1    -3    -4
  01   |     3     4     0    -1
  11   |     5     6     2     1
  10   |     2     3    -1    -2
> print /bit F
 +-: !a & c & !d | !b & c
  2: a & b & !c | !a & c & !d | b & !c & d | !b & c
  1: a & !b | a & d | !a & b & !d
  0: b & !d | !b & d
> print F > 0
 a & b | a & !c | b & !c | !c & d
> M = UpperBound(F)
> print M
     6
> print F == M
 a & b & !c & d
> C = (F >= -1) & (F < 4)
> print C
 a & c & d | !a & !c & !d | b & c | !b & !c
> print /map C
 a b : c d
       |    00    01    11    10
  00   |     1     1     0     0
  01   |     1     0     1     1
  11   |     0     0     1     1
  10   |     1     1     1     0
> quit
%
```

Figure 9.9 An example of executing BEM-II.

compose scripts for various applications much more easily than we can by developing and tuning a specific program.

9.3.2 Subset-Sum Problem

The *subset-sum problem* is one example of a combinatorial problem that can easily be described by arithmetic Boolean expressions and solved by BEM-II. This problem is to find a subset of a given set of positive integers $\{a_1, a_2, a_2, \ldots, a_n\}$, whose sum is a given number b. It is a fundamental problem for many applications, including VLSI CAD systems.

In BEM-II script, we use n input variables for representing whether or not the i-th number is chosen, and the constraint on these input variables can be described with simple arithmetic Boolean expressions. The following is an example of BEM-II script for a subset-sum problem:

```
symbol x1 x2 x3 x4 x5
S = 2*x1 + 3*x2 + 3*x3 + 4*x4 + 5*x5
C = (S == 12)
```

The `C` represents the assignments of the input variables for satisfying the constraint. This expression is almost the same as the definition of the problem. We can easily write and understand this script.

BEM-II is convenient not only for solving the problem but also for analyzing the nature of the problem. We can analyze the behavior of the constraint functions by displaying with various formats. An example of BEM-II execution for a subset-sum problem is shown in Fig. 9.10.

9.3.3 8-Queens Problem

8-queens problem is another good example showing that BEM-II is very convenient to describe the problems.

We first allocate 64 input variables corresponding to the squares on a chessboard. These represent whether or not there is a queen on that square. The constraints that the input variables should satisfy are expressed as follows:

- The sum of eight variables in the same column is 1.
- The sum of eight variables in the same row is 1.
- The sum of variables on the same diagonal line is less than 2.

These constraints can be described with simple arithmetic Boolean expressions as:

```
F1 = (x11 + x12 + x13 + ... + x18 == 1 )
```

```
% bemII
***** Arithmetic Boolean Expression Manipulator (Ver. 4.2) *****
> symbol x1 x2 x3 x4 x5
> S = 2*x1 + 3*x2 + 3*x3 + 4*x4 + 5*x5
> print /map S
 x1 x2 : x3 x4 x5
       |  000   001   011   010 |  110   111   101   100
   00  |    0     5     9     4 |    7    12     8     3
   01  |    3     8    12     7 |   10    15    11     6
   11  |    5    10    14     9 |   12    17    13     8
   10  |    2     7    11     6 |    9    14    10     5
> C = (S == 12)
> print C
 x1 & x2 & x3 & x4 & !x5 | !x1 & x2 & !x3 & x4 & x5 |
      !x1 & !x2 & x3 & x4 & x5
> print /map C ? S : 0
 x1 x2 : x3 x4 x5
       |  000   001   011   010 |  110   111   101   100
   00  |    0     0     0     0 |    0    12     0     0
   01  |    0     0    12     0 |    0     0     0     0
   11  |    0     0     0     0 |   12     0     0     0
   10  |    0     0     0     0 |    0     0     0     0
> print /map (S >= 12)? S : 0
 x1 x2 : x3 x4 x5
       |  000   001   011   010 |  110   111   101   100
   00  |    0     0     0     0 |    0    12     0     0
   01  |    0     0    12     0 |    0    15     0     0
   11  |    0     0    14     0 |   12    17    13     0
   10  |    0     0     0     0 |    0    14     0     0
> quit
%
```

Figure 9.10 Execution of BEM-II for a subset-sum problem.

```
F2 = (x21 + x22 + x23 + ... + x28 == 1 )
...
C = F1 & F2 & ...
```

BEM-II analyzes the above expressions directly. This is much easier than creating a specific program in a programming language. The script for the 8-queens problem took only ten minutes to create.

Table 9.2 shows the results when we applied this method to N-queens problems. In our experiments, we solved the problem up to $N = 11$. When seeking only one solution, we can solve the problem for larger values of N by using a conventional algorithm based on backtracking. However, the conventional method does not enumerate all the

Table 9.2 Results for N-queens problems.

N	Variables	BDD nodes	Solutions	Time(s)
8	64	2450	92	6.1
9	81	9556	352	18.3
10	100	25944	724	68.8
11	121	94821	2680	1081.9

solutions nor does it count the number of solutions for larger Ns. The BDD-based method generates all the solutions simultaneously and keeps them in a BDD. Therefore, if an additional constraint is appended later, we can revise the script easily without rewriting the program from the beginning. This customizability makes BEM-II very efficient in terms of the total time for programming and execution.

9.3.4 Traveling Salesman Problem

Traveling salesman problem (TSP) can also be solved by using BEM-II. The problem is to find the minimum-cost path for visiting all given cities once and returning to the starting city.

If n is the total number of cities, we allocate $n(n-1)/2$ input variables from x_{12} to $x_{(n-1)n}$, where x_{ij} represents the path between i-th city and j-th city. Using this variable scheme, we can express the constraints as follows.

- Each city has two path (coming in and going out):
  $x_{12} + x_{13} + x_{14} + \ldots + x_{1n} = 2$
  $x_{12} + x_{23} + x_{24} + \ldots + x_{2n} = 2$
  ...
  $x_{1n} + x_{2n} + \ldots + x_{(n-1)n} = 2$

- All the cities are connected:

  step-1
  Let $F_1 = 1, \quad F_2, F_3, \ldots, F_n = 0.$

  Repeat step-2 n times.

  step-2
  Let $F_1 = F_1 \vee x_{12}F_2 \vee x_{13}F_3 \vee \ldots \vee x_{1n}F_n.$

Table 9.3 Results for traveling salesman problems.

n	Solutions	BDD nodes	Time(s)
8	2520	2054	8.7
9	66136	20160	28.8
10	181440	19972	216.5

Let $F_2 = F_2 \vee x_{12}F_1 \vee x_{23}F_3 \vee \ldots \vee x_{2n}F_n$.

...

Let $F_n = F_n \vee x_{1n}F_1 \vee x_{2n}F_2 \vee \ldots \vee x_{(n-1)n}F_{n-1}$.

step-3 Condition $C = F_1 \wedge F_2 \wedge \ldots \wedge F_n$.

The logical product of all the above constraint expressions becomes the solution. BEM-II feeds these expressions directly, and generates BDDs representing all the possible paths to visit n cities. As mentioned in previous section, we can specify the cost (distance) of each path, and find an optimal solution to the problem after generating BDDs. Experimental results are listed in Table 9.3. We can solve the problem for n up to 10. This seems poor because the conventional method solves the problem when n is more than 1000. However, our method computes all the solutions at once, and the additional constraints can be specified flexibly. For example,

- There is a path that should be used, or should not be used.
- There is a city that should be visited first (second, third, ...).
- The starting city and the ending city are different.

These constraints can easily be expressed by arithmetic Boolean expressions, and BEM-II feeds them directly to solve the modified problems. It is not too late to develop the application-specific program after trying BEM-II.

9.3.5 Timing Analysis for Logic Circuits

In designing high-speed digital systems, timing analysis for logic circuits is important. The orthodox approach is to traverse the circuit to find the active path with the topologically maximum length. Takahara[Tak93] proposed a new method using

Table 9.4 Results of timing analysis.

Circuit	In.	Out.	Gates	Number of BDD nodes	
				Timing data	Logic data
cm138a	6	8	29	235	129
sel8	12	2	43	926	268
alu2	10	6	434	16,883	4,076
alu4	14	8	809	97,318	9,326
alupla	25	5	114	22,659	2,889
mult6	12	12	411	57,777	9,496
too_large	39	3	1044	730,076	10,789
C432	36	7	255	1,689,576	10,827

BEM-II. This method calculates B-to-I functions representing the delay time with respect to the values of the primary inputs. Using this method, we can completely analyze the timing behavior of a circuit for any combination of input values.

The B-to-I functions for the delay time can be described by a number of arithmetic Boolean expressions, each of which specifies the signal propagation at each gate. For a two-input AND gate with delay D, for example, if T_a and T_b are the signal arrival times at the two input pins, and V_a and V_b are their final logic values, the signal arrival time at output pin T_c is expressed as:

$$T_c = T_b + D \text{ when } (T_a \leq T_b) \text{ and } (V_a = 1),$$
$$T_c = T_a + D \text{ when } (T_a \leq T_b) \text{ and } (V_a = 0),$$
$$T_c = T_a + D \text{ when } (T_a > T_b) \text{ and } (V_b = 1),$$
$$T_c = T_b + D \text{ when } (T_a > T_b) \text{ and } (V_b = 0).$$

These rules can be described by the following arithmetic Boolean expression:

```
Tc = D + ((Ta > Tb)? (Vb? Ta:Tb):(Va? Tb:Ta)).
```

By calculating such expressions for all the gates in the circuit, we can generate BDD vectors for the B-to-I functions of the delay time. Experimental results for practical benchmark circuits are listed in Table 9.4[Tak93]. The size of the BDDs for the delay time is about 20 times greater than that of the BDDs for the Boolean functions of the circuits.

The generated BDDs maintain the timing information for all of the internal nets in the circuit, and we can use BEM-II to analyze the circuits in various ways. For example, we can easily compare the delay times of two nets in the circuit.

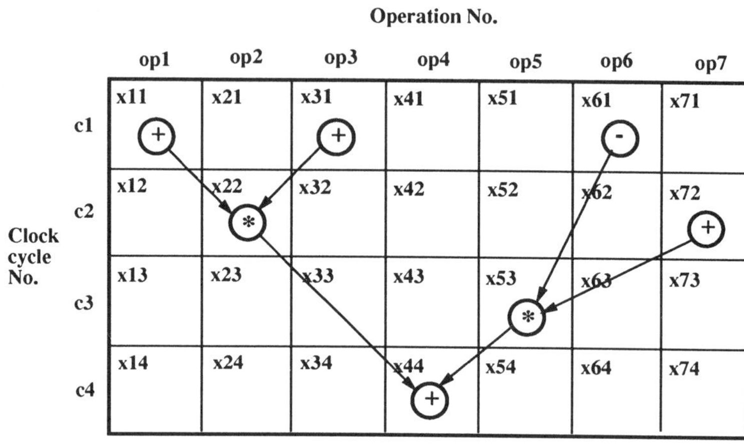

Figure 9.11 Example of a data-flow graph.

9.3.6 Scheduling Problem in Data Path Synthesis

Scheduling is one of the most important subtasks that must be solved to perform data path synthesis. Miyazaki[Miy92] proposed a method for solving scheduling problems by using BEM-II. The problem is to find the minimum-cost scheduling for a procedure specified by a data-flow graph under such constraints as the number of operation units and the maximum number of clock cycles (Fig. 9.11). This problem can be solved by using linear programming, but BEM-II can also be used.

When m is the total number of operations that appear in the data-flow graph and n is the maximum number of clock cycles, we allocate $m \times n$ input variables from x_{11} to x_{mn}, where x_{ij} represents the i-th operation executed in the j-th clock cycle. Using this variable coding, we can represent the constraints of the scheduling problem as follows.

1. Each operation has to be executed once:
   $x_{11} + x_{12} + \ldots + x_{1n} = 1$
   $x_{21} + x_{22} + \ldots + x_{2n} = 1$
   $\ldots$
   $x_{m1} + x_{m2} + \ldots + x_{mn} = 1$

Table 9.5 Results of scheduling problem.

Data	All solutions	Optimal solusions	c-step	Multi-plier	ALUs	BDD nodes	Time (sec)
DiffEq	108	3	4	2	2	321	2.1
Tseng	12	4	5	0	3	321	1.3
EWF	4200	167	14	2	3	1261	27.0

2. The same kind of operations cannot be executed simultaneously beyond the number of operation units. For example, when there are two adders and the a-th, b-th, and c-th operations require an adder:
   $x_{a1} + x_{b1} + x_{c1} \leq 2$
   $x_{a2} + x_{b2} + x_{c2} \leq 2$
   $\ldots$
   $x_{an} + x_{bn} + x_{cn} \leq 2.$

3. If two operations have a dependency in the data-flow graph, the operation in the upper stream has to be executed before the one in the lower stream.
   Let $C_1 = 1 \times x_{11} + 2 \times x_{12} + \ldots + n \times x_{1n}$.
   Let $C_2 = 1 \times x_{21} + 2 \times x_{22} + \ldots + n \times x_{2n}$.
   $\ldots$
   Let $C_m = 1 \times x_{m1} + 2 \times x_{m2} + \ldots + n \times x_{mn}$.
   Then $(C_i < C_j)$ is the condition that the i-th operation is executed before j-th one.

The logical product of all these constraint expressions becomes the solution to the scheduling problem. Using BEM-II, we can easily specify the cost of operation and the other constraints. BEM-II analyzes the above expressions and tries to generate BDDs that represent the solutions. If it succeeds in generating BDDs in main memory, we can immediately find a solution to the problem and count the number of solutions. Table 9.5[Miy92] lists the experimental results for benchmark data from the High-Level Synthesis Workshop (HLSW). The BDDs for constraint functions can be generated in a feasible memory and space.

9.4 CONCLUSION

We have developed an arithmetic Boolean expression manipulator (BEM-II) that can easily solve many kind of combinatorial problems by using arithmetic Boolean expressions. This manipulator can directly computes the equalities and inequalities in the constraints and costs of the problem, and generates BDDs that represent the solutions. It is therefore not necessary to write a specific program in a programming language for solving a problem. In addition to being used in the examples we have shown here, BEM-II can also be used for solving minimum-tree problems, magic squares, crypt-arithmetic problems, etc. Although the computation speed of BEM-II is not as great as that of well-optimized heuristic algorithms for large-scale problems, the customizability of BEM-II makes it very efficient in terms of total time for programming and execution. We expect it to be a useful tool in digital systems research and development.

10 CONCLUSIONS

In this book, we have discussed techniques related to BDDs and their applications for VLSI CAD systems.

In Chapter 2, we presented basic algorithms of Boolean function manipulation using BDDs. We then described implementation techniques to make BDD manipulators applicable to practical problems. As an improvement of BDDs, we proposed the use of *attributed edges*, which are edges attached with several sorts of attributes representing a certain operation. The *negative edge* is particularly effective and now widely used. We implemented a BDD subroutine package for Boolean function manipulation. It can efficiently represent and manipulate very large-scale BDDs containing more than a million of nodes. These techniques have been developed and improved in many laboratories all over the world[Bry86, MIY90, MB88, BRB90], and some program packages are open to the public. They are now widely utilized in research and development on VLSI CAD and other aspects of computer science.

The variable ordering method is one of the most important issues in the application of BDDs. In Chapter 3, we discussed the properties of variable ordering and showed two heuristic methods: *DWA method* and *minimum-width method*. The DWA method finds an appropriate order before generating BDDs. It refers to topological information which specifies the sequence of logic operations in the Boolean expression or logic circuit. Experimental results show that for many practical circuits the DWA method finds a tolerable order in a short computation time. The minimum-width method, on the other hand, finds an appropriate order for a given BDD without using additional information. It seeks a good order from a global viewpoint, rather than by doing an incremental search. In many cases, this method gives a better order than the DWA method does in a longer but still reasonable computation time. The techniques of variable ordering are still being studied intensively, but it is almost impossible to have a method that always finds best order within a practical time. We will therefore have to use some heuristic methods selected according to the application.

In many applications, we sometimes use ternary-valued functions containing *don't cares*. The technique of handling *don't cares* are basic and important for Boolean function manipulation. In Chapter 4, we showed two methods — ternary-valued BDDs and BDD pairs — that we compared by introducing the D-variable. In this discussion, we showed that both the two methods are the special forms of the BDDs using the D-variable and that we can compare the efficiency of the two methods by considering of the properties of variable ordering.

We extended the technique of handling *don't cares* to be used for Binary-to-integer functions, and we presented two methods: MTBDDs and BDD vectors. These method can be compared by introducing the *bit-selection variables*, similarly to the D-variable for the ternary-valued functions. Multi-valued logic manipulation is important to broaden the scope of BDD application. These techniques are useful in various areas of computer science as well as for VLSI CAD.

In Chapter 5, we discussed the topic of transforming a BDD representation into another kind of data structure. We presented the *ISOP algorithm* that generates prime-irredundant cube sets directly from given BDDs, in contrast to the conventional cube set reduction algorithms, which temporarily generate redundant cube sets or truth tables. The experimental results demonstrate that our method is efficient in terms of time and space. In practical time, we can generate cube sets consisting of more than 1,000,000 literals from multi-level logic circuits that have never before been flattened into two-level logics. In experiments with large-scale examples, our method is more than 10 times faster than conventional methods. It gives quasi-minimum numbers of cubes and literals. In terms of size of the result, the ISOP algorithm may give somewhat larger results than ESPRESSO, but there are many applications in which such an increase is tolerable. Our method can be utilized to transform BDDs into compact cube sets or to flatten multi-level circuits into two-level circuits.

As our understanding of BDDs has deepened, their range of applications has broadened. In VLSI CAD problems, we are often faced with manipulating not only Boolean functions but also *sets of combinations*. In Chapter 6, we proposed *Zero-Suppressed BDDs (ZBDDs)*, which are BDDs based on a new reduction rule. ZBDDs enable us to manipulate sets of combinations more efficiently than we can when using conventional BDDs. The effect of ZBDDs is especially remarkable when we are manipulating sparse combinations. We discussed the properties of ZBDDs and their efficiency as revealed by a statistical experiment. We then presented the basic operators for ZBDDs. These operators are defined as the operations on sets of combinations, and differ slightly from the operations on Boolean functions based on conventional BDDs.

On the basis of these ZBDD techniques, we discussed the calculation of unate cube set algebra. We developed efficient algorithms for computing unate cube set operations including multiplication and division, and we showed some practical applications. For solving some types of combinatorial problems, unate cube set algebra is more useful than using conventional logic operations. We expect the unate cube set calculator to be a helpful tool in developing VLSI CAD systems and in various other applications.

In Chapter 7, we presented an application for VLSI logic synthesis. We proposed a fast factorization method for cube set representation based on ZBDDs. Our new algorithm can be executed in a time almost proportional to the number of ZBDD nodes, which is usually much smaller than the number of literals in the cube set. By using this method, we can quickly generate multi-level logics from implicit cube sets even for parity functions and full-adders, which have never been practicable before. We implemented a new multi-level logic synthesizer, and experimental results indicate that our method is much faster than conventional methods, and the difference is remarkable when large cube sets are factorized. Our method greatly accelerates multi-level logic synthesis systems and makes them applicable to larger circuits. There is still some room to improve the results. We have used a simple strategy for choosing divisors, but more sophisticated strategies might be possible. Moreover, a Boolean division method for implicit cube sets is worth investigating to improve the optimization quality.

In Chapter 8, we discussed a ZBDD-based method of manipulating polynomials. We proposed an elegant way to represent polynomials by using ZBDDs, and showed efficient algorithms for operations on those representations. Our experimental results indicate that we can manipulate large-scale polynomials implicitly within a feasible time and space. An important feature of our representation is that it is the canonical form of a polynomial under a fixed variable order. As the polynomial calculus is a basic part of mathematics, our method is very useful in various areas.

In Chapter 9, we described a useful tool for the research on computer science. Using the BDD techniques, we have developed an arithmetic Boolean expression manipulator (BEM-II) that can easily solve many kind of combinatorial problems. It calculates not only binary logic operation but also arithmetic operations on multi-valued logics: such as addition, subtraction, multiplication, division, equality, and inequality. Such arithmetic operations provide simple descriptions for various problems. BEM-II feeds and computes the problems described as equalities and inequalities, which are dealt with using 0-1 linear programming. It is therefore not necessary to write a specific program in a programming language for solving a problem. In this chapter, we discussed the data structure and algorithms for the arithmetic operations. We then presented the specification of BEM-II and some application examples. Experimental results indicate that the customizability of BEM-II makes it very efficient in terms of total time for programming and execution. We therefore expect it to be a useful tool for in digital systems research and development.

A number of works based on the BDD techniques are in progress. The novelty of BDD manipulation are summarized as:

1. Extracts the redundancy contained in the Boolean functions by using a fixed variable order.
2. Eliminates redundancy by ensuring that there are no duplicate nodes and that equivalent computation is not repeated.

The BDD algorithms are based on the quick search of the hash tables and the linked list data structure, both of which greatly benefit from the *random access machine model*,such that any data in the main memory can be accessed in a constant time. Because most computers are designed according to this model, we conclude that the BDD manipulation algorithms are fairly sophisticated and adapted to the conventional computer model. BDDs and their improvements will become key techniques in VLSI CAD systems and other aspects of computer science.

REFERENCES

[Ake78] S. B. Akers. Binary decision diagram. *IEEE Trans. on Computers*, Vol. C-27, No. 6, pp. 509–516, June 1978.

[BC95] R. E. Bryant and Y.-A. Chen. Verification of arithmetic circuits with binary moment diagrams. In *Proc. of 32nd ACM/IEEE Design Automation Conference (DAC'95)*, pp. 535–541, June 1995.

[BCMD90] J. R. Burch, E. M. Clarke, K. L. McMillan, and D. L. Dill. Sequential circuit verification using symbolic model checking. In *Proc. of 27th ACM/IEEE Design Automation Conference (DAC'90)*, pp. 46–51, June 1990.

[BF85] F. Brglez and H. Fujiwara. A neutral netlist of 10 combinational circuits. In *Proc. of 1985 IEEE International Symposium Circuit and Systems (ISCAS'85), Special Session on ATPG and Fault Simulation*, June 1985.

[BFG+93] R. I. Bahar, E. A. Frohm, C. M. Gaona, G. D. Hachtel, E. Macii, A. Pardo, and F. Somenzi. Algebraic decision diagrams and their applications. In *Proc. of IEEE/ACM International Conference on Computer-Aided Design (ICCAD-93)*, pp. 188–191, November 1993.

[BHMSV84] R. K. Brayton, G. D. Hachtel, C. T. McMullen, and A. L. Sangiovanni-Vincentelli. *Logic Minimization Algorithms for VLSI Synthesis*. Kluwer Academic Publishers USA, 1984.

[BRB90] K. S. Brace, R. L. Rudell, and R. E. Bryant. Efficient implementation of a BDD package. In *Proc. of 27th ACM/IEEE Design Automation Conference (DAC'90)*, pp. 40–45, June 1990.

[BRKM91] K. M. Butler, D. E. Ross, R. Kapur, and M. R. Mercer. Heuristics to compute variable orderings for efficient manipulation of ordered binary decision diagrams. In *Proc. of 28th ACM/IEEE Design Automation Conference (DAC'91)*, pp. 417–420, June 1991.

[Bry86] R. E. Bryant. Graph-based algorithms for Boolean function manipulation. *IEEE Trans. on Computers*, Vol. C-35, No. 8, pp. 677–691, August 1986.

[Bry91] R. E. Bryant. On the complexity of VLSI implementations and graph representations of Boolean functions with application to integer multiplication. *IEEE Trans. on Computers*, Vol. C-40, No. 2, pp. 205–213, February 1991.

[BSVW87] R. K. Brayton, R. R. A. Sangiovanni-Vincentelli, and A. R. Wang. MIS: A multiple-level logic optimization system. *IEEE Trans. on Computer-Aided Design*, Vol. CAD-6, No. 6, pp. 1062–1081, November 1987.

[CB89] K. Cho and R. E. Bryant. Test pattern generation for sequential MOS circuits by symbolic fault simulation. In *Proc. of 26th ACM/IEEE Design Automation Conference (DAC'89)*, pp. 418–423, June 1989.

[CHJ+90] H. Cho, G. D. Hachtel, S.-W. Jeong, B. Plessier, E. Schwarz, and F. Somenzi. ATPG aspects of FSM verification. In *Proc. of IEEE/ACM International Conference on Computer-Aided Design (ICCAD-90)*, pp. 134–137, November 1990.

[CM92] O. Coudert and J. C. Madre. A new graph based prime computation technique. In T. Sasao, editor, *New Trends in Logic Synthesis*, chapter 2, pp. 33–57. Kluwer Academic Publishers USA, 1992.

[CMF93] O. Coudert, J. C. Madre, and H. Fraisse. A new viewpoint of two-level logic optimization. In *Proc. of 30th ACM/IEEE Design Automation Conference (DAC'93)*, pp. 625–630, June 1993.

[CMZ+93] E. M. Clarke, K. L. McMillan, X. Zhao, M. Fujita, and J. Yang. Spectral transforms for large Boolean functions with applications to technology mapping. In *Proc. of 30th ACM/IEEE Design Automation Conference (DAC'93)*, pp. 54–60, June 1993.

[dG86] A. J. de Geus. Logic synthesis and optimization benchmarks for the 1986 DAC. In *Proc. of 23rd ACM/IEEE Design Automation Conference (DAC'86)*, pp. 78, June 1986.

[FFK88] M. Fujita, H. Fujisawa, and N. Kawato. Evaluation and improvement of Boolean comparison method based on binary decision diagrams. In *Proc. of IEEE/ACM International Conference on Computer-Aided Design (ICCAD-88)*, pp. 2–5, November 1988.

[FMK91] M. Fujita, Y. Matsunaga, and T. Kakuda. On variable ordering of binary decision diagrams for the application of multi-level logic synthesis. In *Proc. of IEEE The European Conference on Design Automation (EDAC'91)*, pp. 50–54, February 1991.

[FS87] S. J. Friedman and K. J. Supowit. Finding the optimal variable ordering for binary decision diagrams. In *Proc. of 24th ACM/IEEE Design Automation Conference (DAC'87)*, pp. 348–356, June 1987.

[HCO74] S. J. Hong, R. G. Cain, and D. L. Ostapko. MINI: A heuristic approach for logic minimization. *IBM Journal of Research and Development*, Vol. 18, No. 5, pp. 443–458, 1974.

[HMPS94] G. D. Hachtel, E. Macii, A. Pardo, and F. Somenzi. Probabilistic analysis of large finite state machines. In *Proc. of 31st ACM/IEEE Design Automation Conference (DAC'94)*, pp. 270–275, June 1994.

[Ish92] N. Ishiura. Synthesis of multi-level logic circuits from binary decision diagrams. In *Proc. of Synthesis and Simulation Meeting and International Interchange (SASIMI'92, Japan)*, pp. 74–83, March 1992.

[ISY91] N. Ishiura, H. Sawada, and S. Yajima. Minimization of binary decision diagrams based on exchanges of variables. In *Proc. of IEEE/ACM International Conference on Computer-Aided Design (ICCAD-91)*, pp. 472–475, November 1991.

[IY90] N. Ishiura and S. Yajima. A class of logic functions expressible by a polynomial-size binary decision diagrams. In *Proc. of Synthesis and Simulation Meeting and International Interchange (SASIMI'90, Japan)*, pp. 48–54, October 1990.

[IYY87] N. Ishiura, H. Yasuura, and S. Yajima. High-speed logic simulation on vector processors. *IEEE Trans. on Computer-Aided Design*, Vol. CAD-6, No. 3, pp. 305–321, May 1987.

[JT92] R. P. Jacobi and A. M. Trullemans. Generating prime and irredundant covers for binary decision diagrams. In *Proc. of IEEE The European Conference on Design Automation (EDAC'92)*, pp. 104–108, March 1992.

[LCM89] P. Lammens, L. J. Claesen, and H. D. Man. Tautology checking benchmarks: Results with TC. In *Proc. of IFIP International Workshop on Applied Formal Methods for Correct VLSI Design*, pp. 600–604, November 1989.

[LL92] H.-T. Liaw and C.-S. Lin. On the OBDD-representation of general Boolean functions. *IEEE Trans. on Computers*, Vol. C-41, No. 6, pp. 661–664, June 1992.

[LPV93] Y.-T. Lai, M. Pedram, and S. B. Vrudhula. FGILP: An integer linear program solver based on function graphs. In *Proc. of IEEE/ACM International Conference on Computer-Aided Design (ICCAD-93)*, pp. 685–689, November 1993.

[LS90] B. Lin and F. Somenzi. Minimization of symbolic relations. In *Proc. of IEEE/ACM International Conference on Computer-Aided Design (ICCAD-90)*, pp. 88–91, November 1990.

[MB88] J. C. Madre and J. P. Billon. Proving circuit correctness using formal comparison between expected and extracted behaviour. In *Proc. of 25th ACM/IEEE Design Automation Conference (DAC'88)*, pp. 205–210, June 1988.

[MF89] Y. Matsunaga and M. Fujita. Multi-level logic optimization using binary decision diagrams. In *Proc. of IEEE/ACM International Conference on Computer-Aided Design (ICCAD-89)*, pp. 556–559, November 1989.

[Min92a] S. Minato. Fast generation of irredundant sum-of-products forms from binary decision diagrams. In *Proc. of Synthesis and Simulation Meeting and International Interchange (SASIMI'92, Japan)*, pp. 64–73, April 1992.

[Min92b] S. Minato. Minimum-width method of variable ordering for binary decision diagrams. *IEICE Trans. Fundamentals*, Vol. E75-A, No. 3, pp. 392–399, March 1992.

[Min93a] S. Minato. BEM-II: An arithmetic Boolean expression manipulator using BDDs. *IEICE Trans. Fundamentals*, Vol. E76-A, No. 10, pp. 1721–1729, October 1993.

[Min93b] S. Minato. Fast generation of prime-irredundant covers from binary decision diagrams. *IEICE Trans. Fundamentals*, Vol. E76-A, No. 6, pp. 967–973, June 1993.

[Min93c] S. Minato. Fast weak-division method for implicit cube representation. In *Proc. of Synthesis and Simulation Meeting and International Interchange (SASIMI'93, Japan)*, pp. 423–432, October 1993.

[Min93d] S. Minato. Zero-suppressed BDDs for set manipulation in combinatorial problems. In *Proc. of 30th ACM/IEEE Design Automation Conference (DAC'93)*, pp. 272–277, June 1993.

[Min94] S. Minato. Calculation of unate cube set algebra using zero-suppressed BDDs. In *Proc. of 31st ACM/IEEE Design Automation Conference (DAC'94)*, pp. 420–424, June 1994.

[Min95] S. Minato. Implicit manipulation of polynomials using zero-suppressed BDDs. In *Proc. of IEEE The European Design and Test Conference (ED&TC'95)*, pp. 449–454, March 1995.

[MIY89] S. Minato, N. Ishiura, and S. Yajima. Symbolic simulation using shared binary decision diagram. In *Record of the 1989 IEICE Fall Conference, (in Japanese)*, pp. 1.206–1.207, SA–7–5, September 1989.

[MIY90] S. Minato, N. Ishiura, and S. Yajima. Shared binary decision diagram with attributed edges for efficient Boolean function manipulation. In *Proc. of 27th ACM/IEEE Design Automation Conference (DAC'90)*, pp. 52–57, June 1990.

[Miy92] T. Miyazaki. Boolean-based formulation for data path synthesis. In *Proc. of IEEE Asia-Pacific Conference on Circuits and Systems (APC-CAS'92)*, pp. 201–205, December 1992.

[MKLC87] S. Muroga, Y. Kambayashi, H. C. Lai, and J. N. Culliney. The transduction method—design of logic networks based on permissible functions. *IEEE Trans. on Computers*, Vol. C-38, No. 10, pp. 1404–1424, June 1987.

[Mor70] E. Morreale. Recursive operators for prime implicant and irredundant normal form determination. *IEEE Trans. on Computers*, Vol. C-19, No. 6, pp. 504–509, June 1970.

[MSB93] P. McGeer, J. Sanghavi, and R. K. Brayton. Espresso-signature: A new exact minimizer for logic functions. In *Proc. of 30th ACM/IEEE Design Automation Conference (DAC'93)*, pp. 618–624, June 1993.

[MWBSV88] S. Malik, A. R. Wang, R. K. Brayton, and A. L. Sangiovanni-Vincentelli. Logic verification using binary decision diagrams in a logic synthesis environment. In *Proc. of IEEE/ACM International Conference on Computer-Aided Design (ICCAD-88)*, pp. 6–9, November 1988.

[Oku94] H. G. Okuno. Reducing combinatorial explosions in solving search-type combinatorial problems with binary decision diagram. *Trans. of Information Processing Society of Japan (IPSJ), (in Japanese)*, Vol. 35, No. 5, pp. 739–753, May 1994.

[Rud93] R. Rudell. Dynamic variable ordering for ordered binary decision diagrams. In *Proc. of IEEE/ACM International Conference on Computer-Aided Design (ICCAD-93)*, pp. 42–47, November 1993.

[Tak93] A. Takahara. A timing analysis method for logic circuits. In *Record of the 1993 IEICE Spring Conference (in Japanese)*, pp. 1.120, A–20, March 1993.

[THY93] S. Tani, K. Hamaguchi, and S. Yajima. The complexity of the optimal variable ordering of a shared binary decision diagram. Technical Report 93-6, Department of Information Science, Faculty of Science, University of Tokyo, December 1993.

[TIY91] N. Takahashi, N. Ishiura, and S. Yajima. Fault simulation for multiple faults using BDD representation of fault sets. In *Proc. of IEEE/ACM International Conference on Computer-Aided Design (ICCAD-91)*, pp. 550–553, November 1991.

INDEX